T0253557

Machine Learning: Foundations, Methodologies, and Applications

Series Editors

Kay Chen Tan, Department of Computing, Hong Kong Polytechnic University, Hong Kong, China

Dacheng Tao, University of Technology, Sydney, Australia

Books published in this series focus on the theory and computational foundations, advanced methodologies and practical applications of machine learning, ideally combining mathematically rigorous treatments of a contemporary topics in machine learning with specific illustrations in relevant algorithm designs and demonstrations in real-world applications. The intended readership includes research students and researchers in computer science, computer engineering, electrical engineering, data science, and related areas seeking a convenient medium to track the progresses made in the foundations, methodologies, and applications of machine learning.

Topics considered include all areas of machine learning, including but not limited to:

- Decision tree
- Artificial neural networks
- Kernel learning
- Bayesian learning
- Ensemble methods
- Dimension reduction and metric learning
- Reinforcement learning
- Meta learning and learning to learn
- Imitation learning
- Computational learning theory
- Probabilistic graphical models
- Transfer learning
- Multi-view and multi-task learning
- Graph neural networks
- Generative adversarial networks
- Federated learning

This series includes monographs, introductory and advanced textbooks, and state-of-the-art collections. Furthermore, it supports Open Access publication mode.

More information about this series at https://link.springer.com/bookseries/16715

Teik Toe Teoh • Zheng Rong

Artificial Intelligence with Python

 Springer

Teik Toe Teoh [ORCID]
Nanyang Business School
Nanyang Technological University
Singapore, Singapore

Zheng Rong
Nanyang Technological University
Singapore, Singapore

ISSN 2730-9908 ISSN 2730-9916 (electronic)
Machine Learning: Foundations, Methodologies, and Applications
ISBN 978-981-16-9322-9 ISBN 978-981-16-8615-3 (eBook)
https://doi.org/10.1007/978-981-16-8615-3

This Springer imprint is published by the registered company Springer Nature Singapore Pte Ltd.
The registered company address is: 152 Beach Road, #21-01/04 Gateway East, Singapore 189721,
Singapore

Preface

This book is a practical guide to Python programming and artificial intelligence, written by Dr. Teoh Teik Toe. It contains many articles, notes, and lessons learnt on Python programming, artificial intelligence, and deep learning during Dr. Teoh's career as a deep learning practitioner and a trusted advisor.

Dr. Teoh has been pursuing research in big data, deep learning, cybersecurity, artificial intelligence, machine learning, and software development for more than 25 years. His works have been published in more than 50 journals, conference proceedings, books, and book chapters. His qualifications include a PhD in computer engineering from the NTU, Doctor of Business Administration from the University of Newcastle, Master of Law from the NUS, LLB and LLM from the UoL, and CFA, ACCA and CIMA. He has more than 15 years' experience in data mining, quantitative analysis, data statistics, finance, accounting, and law and is passionate about the synergy between business and technology. He believes that artificial intelligence should be made easy for all to understand and is eager to share his knowledge of the field.

Zheng Rong is a software engineer with 4 years of experience. He embraces the ambiguity of data and enjoys the challenges presented by business problems. He has 3 years of teaching experience in data mining and data science, and coauthored three journal publications on artificial intelligence and deep learning. He is interested in making artificial intelligence programming and technology easy to understand for all, including those from a non-technical background.

The field of artificial intelligence is very broad. It focuses on creating systems capable of executing tasks which would require some form of human intelligence. In-depth knowledge and understanding of the field is required to be able to develop good artificial intelligence programs. The concepts used in self-driving cars and virtual assistants like Amazon's Alexa may seem very complex and difficult to grasp. Entering the field of artificial intelligence and data science can seem daunting to beginners with little to no prior background, especially those with no programming experience.

Throughout his career, Dr. Teoh has delivered many lectures to students from all walks of life about artificial intelligence. There were many students who had limited experience in programming and began with no knowledge of artificial intelligence. However, under his guidance, they eventually gained confidence in writing their own artificial intelligence programs. Through the materials compiled in this book, he hopes to empower more beginners who are eager to study artificial intelligence and enrich their learning process. Hence, the aim of *Artificial Intelligence in Python* is to make AI accessible and easy to understand for people with little to no programming experience through practical exercises. By going through the materials covered in this book, newcomers will gain the knowledge they need on how to create such systems, which are capable of executing tasks that require some form of human-like intelligence.

This book will begin by introducing readers to various topics and examples of programming in Python, as well as key concepts in artificial intelligence. Python will be introduced, and programming skills will be imparted as we go along. Concepts and code snippets will be covered in a step-by-step manner to guide and instill confidence in beginners. Complex subjects in deep learning and machine learning will be broken down into easy-to-digest content and examples. Basics of artificial intelligence, such as classification and regression, will be imparted to build a solid foundation for beginners before moving to more advanced chapters. Artificial intelligence implementations will also be shared, allowing beginners to generate their own artificial intelligence algorithms for reinforcement learning, style transfer, chatbots, and speech and natural language processing.

Singapore, Singapore   Teik Toe Teoh
                                                                  Zheng Rong

Acknowledgments

We would like to acknowledge and thank all our families and friends who have supported us throughout this journey, as well as all those who have helped make this book possible.

Our tutorials and code are compiled from various sources. Without the work of the authors in the references, our book would not have been possible. Credits for Python installation goes to `quantecon` and `pdflatex`. The following codes are compiled so that it can be a quick guide and reference.

Contents

Part I
Python

Chapter 1
Python for Artificial Intelligence

Abstract Python is a very popular programming language with many great features for developing Artificial Intelligence. Many Artificial Intelligence developers all around the world use Python. This chapter will provide an introduction to the Python Programming Language before covering its history and common uses and explain why it is so popular among Artificial Intelligence developers.

Learning outcomes:

- Introduce the Python programming language.

"Python has gotten sufficiently weapons grade that we don't descend into R anymore. Sorry, R people. I used to be one of you but we no longer descend into R." – Chris Wiggins

Python is a general-purpose programming language conceived in 1989 by Dutch programmer Guido van Rossum.

Python is free and open source, with development coordinated through the Python Software Foundation.

Python has experienced rapid adoption in the last decade and is now one of the most commonly used programming languages.

1.1 Common Uses

Python is a general-purpose language used in almost all application domains such as

- Communications
- Web development (Flask and Django covered in the future chapters)
- CGI and graphical user interfaces
- Game development
- AI and data science (very popular)

T. T. Teoh, Z. Rong, *Artificial Intelligence with Python*,
Machine Learning: Foundations, Methodologies, and Applications,
https://doi.org/10.1007/978-981-16-8615-3_1

- Multimedia, data processing, security, etc.

Python is beginner-friendly and routinely used to teach computer science and programming in the top computer science programs.

Python is particularly popular within the scientific and data science communities.

It is steadily replacing familiar tools like Excel in the fields of finance and banking.

1.1.1   Relative Popularity

The following chart, produced using Stack Overflow Trends, shows one measure of the relative popularity of Python.

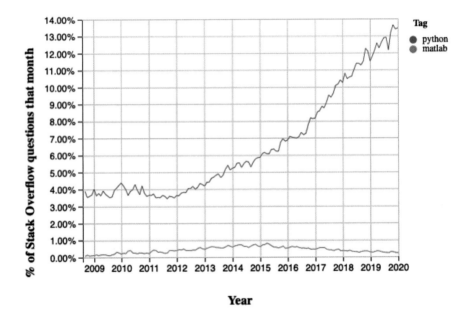

The figure indicates not only that Python is widely used but also that adoption of Python has accelerated significantly since 2012.

This is driven at least in part by uptake in the scientific domain, particularly in rapidly growing fields like data science.

For example, the popularity of pandas, a library for data analysis with Python, has exploded, as seen here.

(The corresponding time path for MATLAB is shown for comparison.)

Note that pandas takes off in 2012, which is the same year that we see Python's popularity begins to spike in the first figure.

Overall, it is clear that

- Python is one of the most popular programming languages worldwide.
- Python is a major tool for scientific computing, accounting for a rapidly rising share of scientific work around the globe.

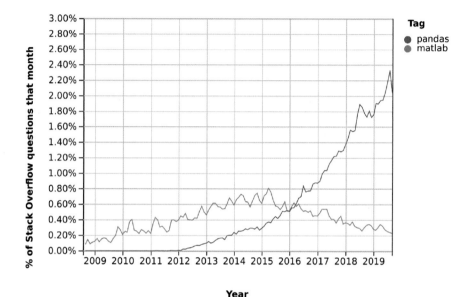

1.1.2   Features

Python is a high-level language suitable for rapid development.

It has a relatively small core language supported by many libraries.

Multiple programming styles are supported (procedural, object-oriented, functional, etc.)

Python is interpreted rather than compiled.

1.1.3   Syntax and Design

One nice feature of Python is its elegant syntax—we will see many examples later on.

Elegant code might sound superfluous, but in fact it is highly beneficial because it makes the syntax easy to read and easy to remember.

Remembering how to read from files, sort dictionaries, and other such routine tasks means that you do not need to break your flow in order to hunt down correct syntax.

Closely related to elegant syntax is an elegant design.

Features like iterators, generators, decorators, and list comprehensions make Python highly expressive, allowing you to get more done with less code.

Namespaces improve productivity by cutting down on bugs and syntax errors.

1.2 Scientific Programming

Python has become one of the core languages of scientific computing.

It is either the dominant player or a major player in

- Machine learning and data science
- Astronomy
- Artificial intelligence
- Chemistry
- Computational biology
- Meteorology

Its popularity in economics is also beginning to rise.

1.3 Why Python for Artificial Intelligence

Python is very popular for Artificial Intelligence developers for a few reasons:

1. It is easy to use:

- Python is easy to use and has a fast learning curve. New data scientists can easily learn Python with its simple to utilize syntax and better comprehensibility.
- Python additionally gives a lot of data mining tools that help in better handling of the data, for example, Rapid Miner, Weka, Orange, and so on.
- Python is significant for data scientists since it has many useful and easy to use libraries like Pandas, NumPy, SciPy, TensorFlow, and many more concepts that a skilled Python programmer must be well acquainted with.

2. Python is flexible:

- Python not only lets you create software but also enables you to deal with the analysis, computing of numeric and logical data, and web development.

- Python has additionally become ubiquitous on the web, controlling various prominent websites with web development frameworks like TurboGears, Django, and Tornado.
- It is perfect for developers who have the talent for application and web development. No big surprise, most data scientists favor this to the next programming alternatives available in the market.

3. Python builds better analytics tools:

- Data analytics is a necessary part of data science. Data analytics tools give information about different frameworks that are important to assess the performance in any business. Python programming language is the best choice for building data analytics tools.
- Python can easily provide better knowledge, get examples, and correlate data from big datasets. Python is additionally significant in self-service analytics. Python has likewise helped the data mining organizations to all the more likely to handle the data for their sake.

4. Python is significant for deep learning:

- Python has a lot of packages like TensorFlow, Keras, and Theano that are assisting data scientists with developing deep learning algorithms. Python gives superior help with regard to deep learning algorithms.
- Deep learning algorithms were inspired by the human brain architecture. It manages to build artificial neural networks that reenact the conduct of the human mind. Deep learning neural networks give weight and biasing to different input parameters and give the desired output.

5. Huge community base:

- Python has a gigantic community base of engineers and data scientists like Python.org, Fullstackpython.com, realpython.com, etc. Python developers can impart their issues and thoughts to the community. Python Package Index is an extraordinary place to explore the different skylines of the Python programming language. Python developers are continually making enhancements in the language that is helping it to turn out to be better over time.

Chapter 2
Getting Started

Abstract In order to write Python code effectively, this chapter will introduce various software development tools that will aid the learning and development process. Jupyter Notebook and Anaconda are very useful tools that will make programming in Python simpler for learners and are commonly used in the industry. Setting up your development environment will be simple to achieve through following our step-by-step guide.

In this chapter, we will be focusing on setting up our Python environment.

2.1 Setting up Your Python Environment

In this lecture, you will learn how to

1. Set up your very own Python environment and get it up and running.
2. Execute simple Python commands.
3. Run a sample program.
4. Install the code libraries that are needed for our lecture.

2.2 Anaconda

The core Python package is easy to install but **not** what you should choose for these lectures.

We require certain libraries and the scientific programing eco system, which

- The core installation (e.g., Python 3.7) does not provide.
- Is painful to install one piece at a time (*concept of a package manager*).

© The Author(s), under exclusive license to Springer Nature Singapore Pte Ltd. 2022   9
T. T. Teoh, Z. Rong, *Artificial Intelligence with Python*,
Machine Learning: Foundations, Methodologies, and Applications,
https://doi.org/10.1007/978-981-16-8615-3_2

Hence, we will be using a distribution with the following features:

1. The core Python language **and**
2. Compatible versions of the most popular scientific libraries.

And, we are using Anaconda.
Anaconda is

- Very popular
- Cross-platform
- Comprehensive
- Completely unrelated to the Nicki Minaj song of the same name

Anaconda also comes with a great package management system to organize your code libraries.

Note All of what follows assumes that you adopt this recommendation.

2.2.1 Installing Anaconda

To install Anaconda, download the binary and follow the instructions.
Important points:

- Install the latest version.
- If you are asked during the installation process whether you would like to make Anaconda your default Python installation, say yes.

2.2.2 Further Installation Steps

Choose the version corresponding to your operating system (Fig. 2.1).
Choose Python 3.6 or higher (for newer versions; Fig. 2.2).

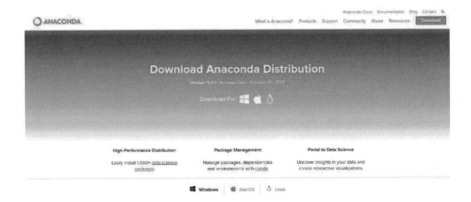

Fig. 2.1 Anaconda installation page

Fig. 2.2 Python version

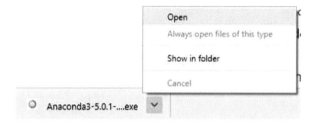

Fig. 2.3 After you have downloaded your installer

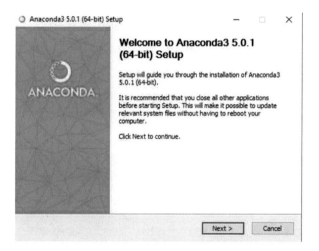

Fig. 2.4 Continue installation

Open the installer after you have downloaded it. Double click on it to open installer (Fig. 2.3).

Click next (Fig. 2.4).

Accept the terms and click next (Fig. 2.5).

Use this setting (Fig. 2.6).

Use start menu and find anaconda prompt (Figs. 2.7 and 2.8).

Congratulations! You have successfully installed anaconda.

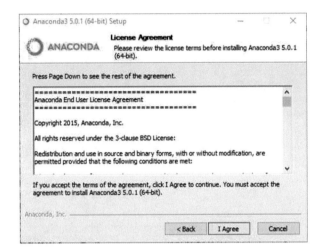

Fig. 2.5 Continue installation

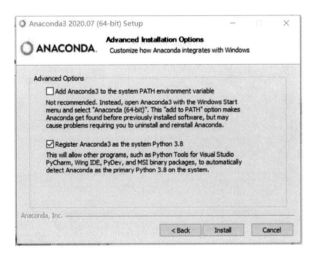

Fig. 2.6 Continue installation

Fig. 2.7 Use start menu and find anaconda prompt

```
Anaconda Prompt (anaconda3_) - python                                                          —

(base) C:\Users\Zheng_>python
Python 3.7.6 (default, Jan  8 2020, 20:23:39) [MSC v.1916 64 bit (AMD64)] :: Anaconda, Inc. on win32
Type "help", "copyright", "credits" or "license" for more information.
>>>
```

Fig. 2.8 Anaconda prompt—used for installing packages

2.2.3   Updating Anaconda

Anaconda supplies a tool called `conda` to manage and upgrade your Anaconda packages.

One `conda` command you should execute regularly is the one that updates the whole Anaconda distribution.

As a practice run, please execute the following:

1. Open up a terminal.
2. Type `conda update anaconda`.

For more information on `conda`, type `conda help` in a terminal.

2.3   Installing Packages

Open up anaconda prompt.

Use start menu and find anaconda prompt (Fig. 2.9).

Command: `pip install \<package-name\>`

2.4   Virtual Environment

A virtual environment is a Python environment such that the Python interpreter, libraries, and scripts installed into it are isolated from those installed in other virtual environment (Fig. 2.10).

Command: `python -m venv \<name of env\>`

Once you have created a virtual environment, you may activate it (Fig. 2.11).

On Windows, run:

`\<name of env\>\Scripts\activate.bat`

On Unix or MacOS, run:

`source \<name of env\>/bin/activate`

After which we can `pip` install any libraries using the command above

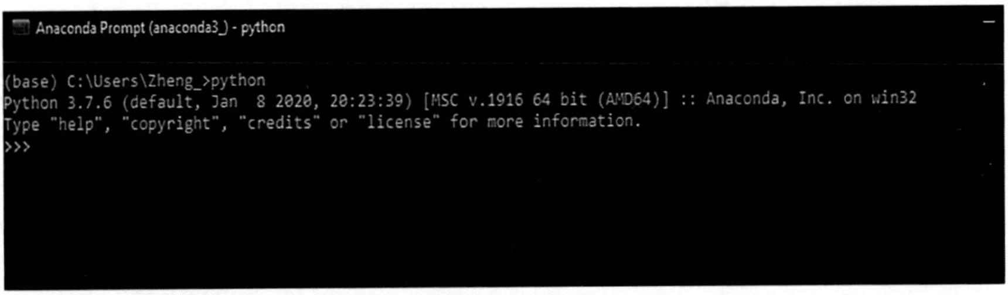

```
Anaconda Prompt (anaconda3_) - python                                                                       —

(base) C:\Users\Zheng_>python
Python 3.7.6 (default, Jan  8 2020, 20:23:39) [MSC v.1916 64 bit (AMD64)] :: Anaconda, Inc. on win32
Type "help", "copyright", "credits" or "license" for more information.
>>>
```

Fig. 2.9 Anaconda prompt—used for installing packages

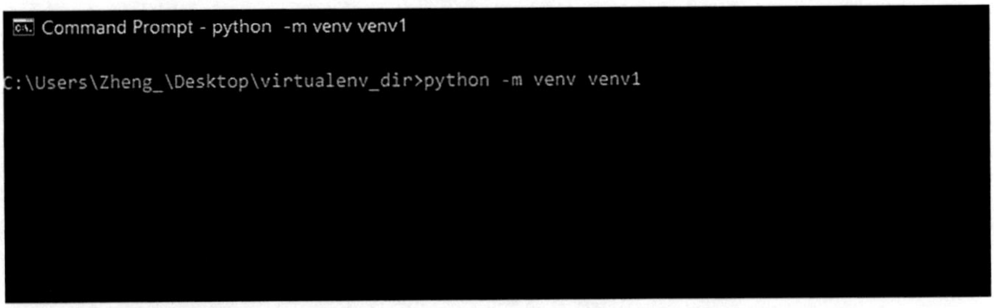

```
Command Prompt - python  -m venv venv1

C:\Users\Zheng_\Desktop\virtualenv_dir>python -m venv venv1
```

Fig. 2.10 Create virtual environment

```
▣ Command Prompt

C:\Users\Zheng_\Desktop\virtualenv_dir>python -m venv venv1

C:\Users\Zheng_\Desktop\virtualenv_dir>venv1\\Scripts\\activate.bat

(venv1) C:\Users\Zheng_\Desktop\virtualenv_dir>
```

Fig. 2.11 Activate virtual environment

2.5 Jupyter Notebooks

Jupyter notebooks are one of the many possible ways to interact with Python and the scientific libraries.

They use a *browser-based* interface to Python with

- The ability to write and execute Python commands.
- Formatted output in the browser, including tables, figures, animation, etc.
- The option to mix in formatted text and mathematical expressions.

Because of these features, Jupyter is now a major player in the scientific computing ecosystem (Fig. 2.12).

While Jupyter is not the only way to code in Python, it is great for when you wish to

- Get started.
- Test new ideas or interact with small pieces of code.
- Share scientific ideas with students or colleagues.

2.5.1 Starting the Jupyter Notebook

Once you have installed Anaconda, you can start the Jupyter Notebook. Either

- Search for Jupyter in your applications menu, or
- Open up a terminal and type `jupyter notebook`.
- Windows users should substitute "Anaconda command prompt" for "terminal" in the previous line.

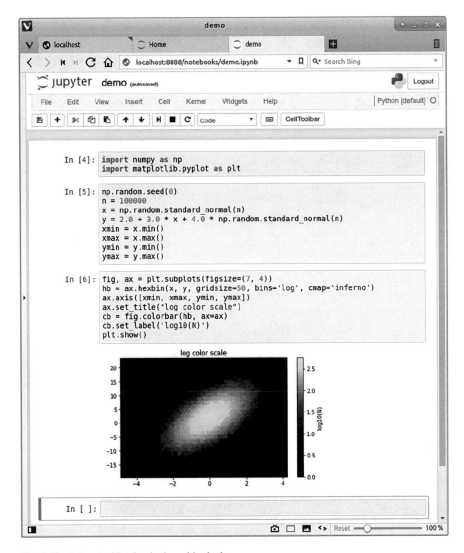

Fig. 2.12 A Jupyter Notebook viewed in the browser

If you use the second option, you will see something like this:

The output tells us the notebook is running at `http://localhost:8888/`.

- `localhost` is the name of the local machine.
- `8888` refers to port number 8888 on your computer.

Thus, the Jupyter kernel is listening for Python commands on port 8888 of our local machine.

Hopefully, your default browser has also opened up with a web page that looks something like this:

What you see here is called the Jupyter *dashboard*.

If you look at the URL at the top, it should be `localhost:8888` or similar, matching the message above.

Assuming all this has worked OK, you can now click on `New` at the top right and select `Python 3` or similar.

Here is what shows up on our machine:

The notebook displays an *active cell*, into which you can type Python commands.

2.5.2 Notebook Basics

Let us start with how to edit code and run simple programs.

Running Cells

Notice that, in the previous figure, the cell is surrounded by a green border.

This means that the cell is in *edit mode*.

In this mode, whatever you type will appear in the cell with the flashing cursor.

When you are ready to execute the code in a cell, hit `Shift-Enter` instead of the usual `Enter`.

(Note: There are also menu and button options for running code in a cell that you can find by exploring.)

Modal Editing

The next thing to understand about the Jupyter Notebook is that it uses a *modal* editing system.

This means that the effect of typing at the keyboard **depends on which mode you are in**.

The two modes are

1. Edit mode

   - It is indicated by a green border around one cell, plus a blinking cursor.
   - Whatever you type appears as is in that cell.

2. Command mode

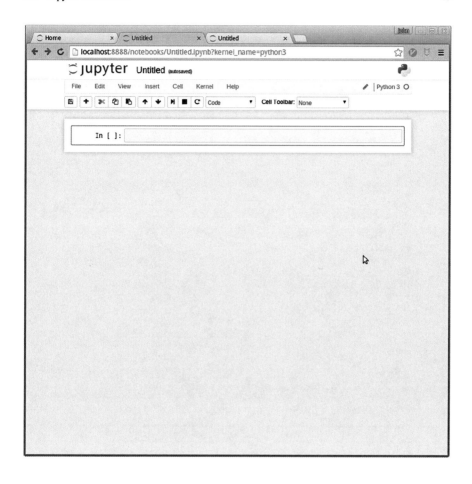

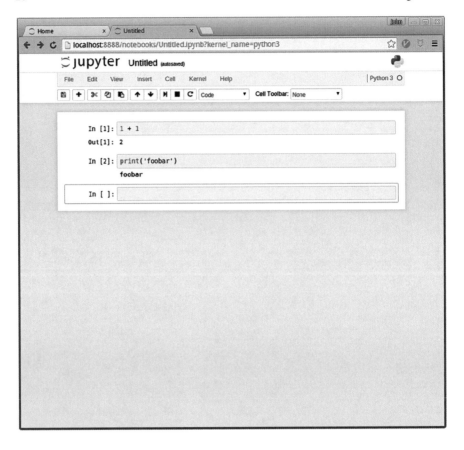

- The green border is replaced by a gray (or gray and blue) border.
- Keystrokes are interpreted as commands—for example, typing b adds a new cell below the current one.

To switch to

- Command mode from edit mode, hit the Esc key or Ctrl-M.
- Edit mode from command mode, hit Enter or click in a cell.

The modal behavior of the Jupyter Notebook is very efficient when you get used to it.

Inserting Unicode (e.g., Greek Letters)

Python supports unicode, allowing the use of characters such as α and β as names in your code.

In a code cell, try typing \alpha and then hitting the tab key on your keyboard.

A Test Program

Let us run a test program.

Here is an arbitrary program we can use: http://matplotlib.org/3.1.1/gallery/pie_and_polar_charts/polar_bar.html.

On that page, you will see the following code:

```
import numpy as np
import matplotlib.pyplot as plt
%matplotlib inline

# Fixing random state for reproducibility
np.random.seed(19680801)

# Compute pie slices
N = 20
θ = np.linspace(0.0, 2 * np.pi, N, endpoint=False)
radii = 10 * np.random.rand(N)
width = np.pi / 4 * np.random.rand(N)
colors = plt.cm.viridis(radii / 10.)

ax = plt.subplot(111, projection='polar')
ax.bar(θ, radii, width=width, bottom=0.0, color=colors, alpha=0.
↪5)

plt.show()
```

Do not worry about the details for now—let us just run it and see what happens. The easiest way to run this code is to copy and paste it into a cell in the notebook. Hopefully, you will get a similar plot.

2.5.3 Working with the Notebook

Here are a few more tips on working with Jupyter Notebooks.

Tab Completion

In the previous program, we executed the line `import numpy as np`.

- NumPy is a numerical library we will work with in depth.

After this import command, functions in NumPy can be accessed with `np.function_name` type syntax.

- For example, try `np.random.randn(3)`.

We can explore these attributes of `np` using the `Tab` key.

For example, here we type np.ran and hit Tab.

Jupyter offers up the two possible completions, random and rank.

In this way, the Tab key shows you available completion options and can reduce the amount of typing required.

On-Line Help

To get help on np, say, we can execute np?. However, do remember to import numpy first.

```
import numpy as np
?np
```

Documentation appears in a split window of the browser, like so:

Clicking on the top right of the lower split closes the on-line help.

Other Content

In addition to executing code, the Jupyter Notebook allows you to embed text, equations, figures, and even videos in the page.

For example, here we enter a mixture of plain text and LaTeX instead of code.

Next, we press the Esc button to enter command mode and then type m to indicate that we are writing Markdown, a markup language similar to (but simpler than) LaTeX.

(You can also use your mouse to select Markdown from the Code drop-down box just below the list of menu items.)

Now we Shift+Enter to produce this:

2.5.4 Sharing Notebooks

Notebook files are just text files structured in JSON and typically ending with .ipynb.

You can share them in the usual way that you share files—or by using web services such as nbviewer.

The notebooks you see on that site are **static** html representations.

To run one, download it as an ipynb file by clicking on the download icon.

Save it somewhere, navigate to it from the Jupyter dashboard, and then run as discussed above.

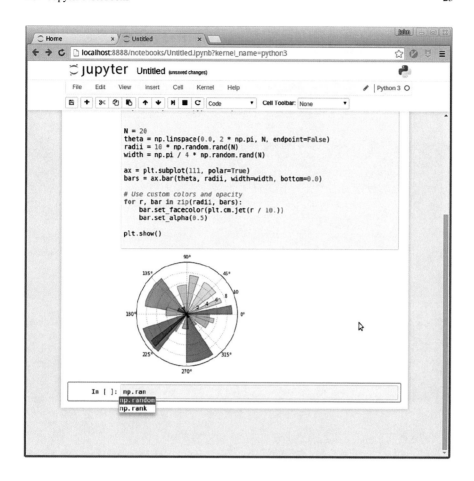

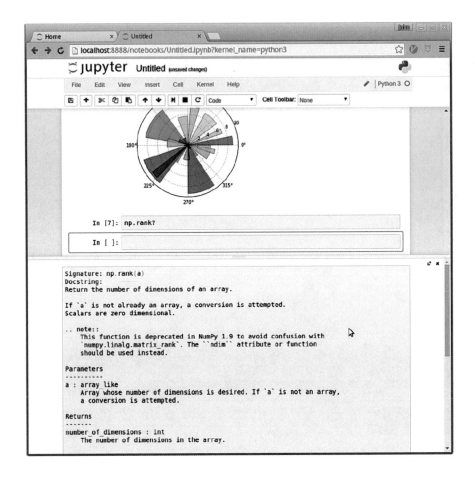

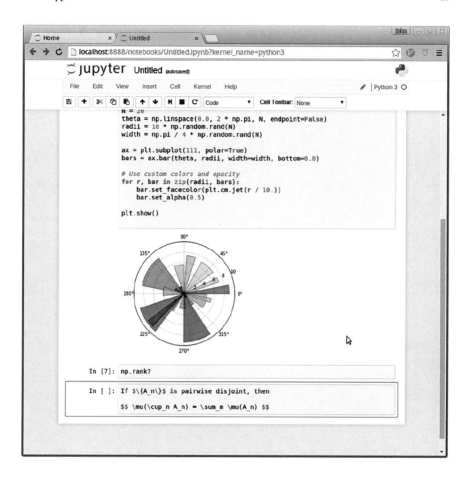

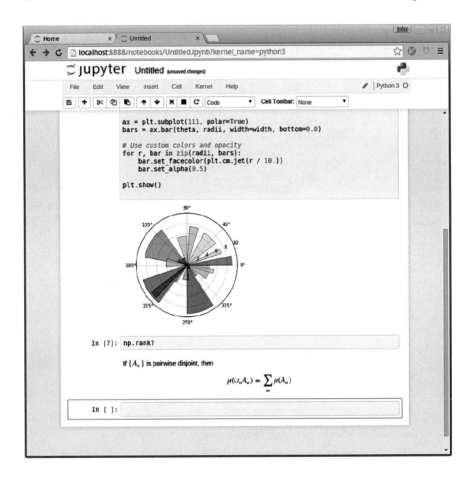

Chapter 3
An Introductory Example

Abstract A few introduction scripts are provided here as we break down the components written in a standard Python script. Introductory concepts such as importing packages, code syntax, and variable names will be covered in this chapter. Basic topics such as lists, loops, and conditions would be touched on to get readers with no Python programming experience started.

Learning outcomes:

- Learn how to import files and packages into Python.
- Learn about lists in Python.
- Learn how to use various loops in Python.

3.1 Overview

We are now ready to start learning the Python language itself.

In this lecture, we will write and then pick apart small Python programs.

The objective is to introduce you to basic Python syntax and data structures.

Deeper concepts will be covered in later lectures.

You should have read the chapter on getting started with Python before beginning this one.

3.2 The Task: Plotting a White Noise Process

Suppose we want to simulate and plot the white noise process $\epsilon_0, \epsilon_1, \ldots, \epsilon_T$, where each draw ϵ_t is independent standard normal.

© The Author(s), under exclusive license to Springer Nature Singapore Pte Ltd. 2022   27
T. T. Teoh, Z. Rong, *Artificial Intelligence with Python*,
Machine Learning: Foundations, Methodologies, and Applications,
https://doi.org/10.1007/978-981-16-8615-3_3

In other words, we want to generate figures that look something like this:
(Here t is on the horizontal axis and ϵ_t is on the vertical axis.)

We will do this in several different ways, each time learning something more about Python.

We run the following command first, which helps ensure that plots appear in the notebook if you run it on your own machine.

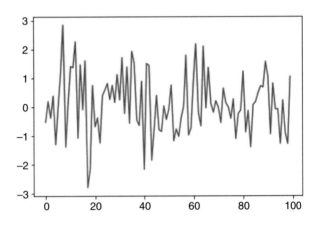

3.3   Our First Program

Here are a few lines of code that perform the task we set:

```
import numpy as np
import matplotlib.pyplot as plt

ε_values = np.random.randn(100)
plt.plot(ε_values)
plt.show()
```

Let us break this program down and see how it works.

3.3.1   *Imports*

The first two lines of the program import functionality from external code libraries.
The first line imports NumPy, a favorite Python package for tasks like

- Working with arrays (vectors and matrices)
- Common mathematical functions like `cos` and `sqrt`
- Generating random numbers

- Linear algebra, etc.

After `import numpy as np`, we have access to these attributes via the syntax `np.attribute`.

Here are two more examples:

```
np.sqrt(4)
```

```
np.log(4)
```

We could also use the following syntax:

```
import numpy

numpy.sqrt(4)
```

But the former method (using the short name `np`) is convenient and more standard.

Why So Many Imports?

Python programs typically require several import statements.

The reason is that the core language is deliberately kept small, so that it is easy to learn and maintain.

When you want to do something interesting with Python, you almost always need to import additional functionality.

Packages

As stated above, NumPy is a Python *package*.

Packages are used by developers to organize code they wish to share.

In fact, a package is just a directory containing

1. Files with Python code—called **modules** in Python speak.
2. Possibly some compiled code that can be accessed by Python (e.g., functions compiled from C or FORTRAN code).
3. A file called __init__.py that specifies what will be executed when we type `import package_name`.

In fact, you can find and explore the directory for NumPy on your computer easily enough if you look around.

On this machine, it is located in

```
anaconda3/lib/python3.7/site-packages/numpy
```

Subpackages

Consider the line ε_values = np.random.randn(100).

Here np refers to the package NumPy, while random is a **subpackage** of NumPy.

Subpackages are just packages that are subdirectories of another package.

3.3.2 Importing Names Directly

Recall this code that we saw above:

```
import numpy as np

np.sqrt(4)
```

Here is another way to access NumPy's square root function

```
from numpy import sqrt

sqrt(4)
```

This is also fine.

The advantage is less typing if we use sqrt often in our code.

The disadvantage is that, in a long program, these two lines might be separated by many other lines.

Then, it is harder for the readers to know where sqrt came from, should they wish to.

3.3.3 Random Draws

Returning to our program that plots white noise, the remaining three lines after the import statements are

```
ε_values = np.random.randn(100)
plt.plot(ε_values)
plt.show()
```

The first line generates 100 (quasi) independent standard normals and stores them in ε_values.

The next two lines generate the plot.

We can and will look at various ways to configure and improve this plot below.

3.4   Alternative Implementations

Let us try writing some alternative versions of our first program, which plotted Independent and Identically Distributed draws from the normal distribution.

The programs below are less efficient than the original one and hence somewhat artificial.

But they do help us illustrate some important Python syntax and semantics in a familiar setting.

3.4.1   A Version with a for Loop

Here is a version that illustrates `for` loops and Python lists.

```python
ts_length = 100
ε_values = []    # empty list

for i in range(ts_length):
    e = np.random.randn()
    ε_values.append(e)

plt.plot(ε_values)
plt.show()
```

In brief,

- The first line sets the desired length of the time series.
- The next line creates an empty *list* called ϵ_values that will store the ϵ_t values as we generate them.
- The statement # empty list is a *comment* and is ignored by Python's interpreter.
- The next three lines are the `for` loop, which repeatedly draws a new random number ϵ_t and appends it to the end of the list ϵ_values.
- The last two lines generate the plot and display it to the user.

Let us study some parts of this program in more detail.

3.4.2   Lists

Consider the statement ϵ_values = [], which creates an empty list.

Lists are a *native Python data structure* used to group a collection of objects.

For example, try

```python
x = [10, 'foo', False]
type(x)
```

The first element of x is an integer, the next is a string, and the third is a Boolean value.

When adding a value to a list, we can use the syntax `list_name.append(some_value)`.

```
x
```

```
x.append(2.5)
x
```

Here `append()` is what is called a *method*, which is a function "attached to" an object—in this case, the list x.

We will learn all about methods later on, but just to give you some idea,

- Python objects such as lists, strings, etc. all have methods that are used to manipulate the data contained in the object.
- String objects have string methods, list objects have list methods, etc.

Another useful list method is `pop()`

```
x
```

```
x.pop()
```

```
x
```

Lists in Python are zero-based (as in C, Java, or Go), so the first element is referenced by `x[0]`.

```
x[0]     # first element of x
```

```
x[1]     # second element of x
```

3.4.3 The for Loop

Now let us consider the `for` loop from the program above, which was

```
for i in range(ts_length):
    e = np.random.randn()
    ε_values.append(e)
```

Python executes the two indented lines `ts_length` times before moving on.

These two lines are called a `code block`, since they comprise the "block" of code that we are looping over.

Unlike most other languages, Python knows the extent of the code block *only from indentation*.

In our program, indentation decreases after line ϵ_values.append(e), telling Python that this line marks the lower limit of the code block.

More on indentation below—for now, let us look at another example of a for loop

```
animals = ['dog', 'cat', 'bird']
for animal in animals:
    print("The plural of " + animal + " is " + animal + "s")
```

This example above helps to illustrate how the for loop works: When we execute a loop of the form

```
for variable_name in sequence:
    <code block>
```

The Python interpreter performs the following:

- For each element of the sequence, it "binds" the name variable_name to that element and then executes the code block.

The sequence object can in fact be a very general object, as we will see soon enough.

3.4.4 A Comment on Indentation

In discussing the for loop, we explained that the code blocks being looped over are delimited by indentation.

In fact, in Python, **all** code blocks (i.e., those occurring inside loops, if clauses, function definitions, etc.) are delimited by indentation.

Thus, unlike most other languages, whitespace in Python code affects the output of the program.

Once you get used to it, this is a good thing. It

- Forces clean, consistent indentation, improving readability.
- Removes clutter, such as the brackets or end statements used in other languages.

On the other hand, it takes a bit of care to get right, so please remember:

- The line before the start of a code block always ends in a colon

  - for i in range(10):
  - if x > y:
  - while x < 100:
  - etc.

- All lines in a code block **must have the same amount of indentation**.
- The Python standard is 4 spaces and that is what you should use.

3.4.5   *While Loops*

The `for` loop is the most common technique for iteration in Python.

But, for the purpose of illustration, let us modify the program above to use a `while` loop instead.

```
ts_length = 100
ε_values = []
i = 0
while i < ts_length:
    e = np.random.randn()
    ε_values.append(e)
    i = i + 1
plt.plot(ε_values)
plt.show()
```

Note that

- The code block for the `while` loop is again delimited only by indentation.
- The statement `i = i + 1` can be replaced by `i += 1`.

3.5   Another Application

Let us do one more application before we turn to exercises.

In this application, we plot the balance of a bank account over time.

There are no withdraws over the time period, the last date of which is denoted by T.

The initial balance is b_0 and the interest rate is r.

The balance updates from period t to $t + 1$ according to

$$b_{t+1} = (1 + r)b_t \tag{3.1}$$

In the code below, we generate and plot the sequence $b_0, b_1, \ldots, b_T$ generated by (3.1).

Instead of using a Python list to store this sequence, we will use a NumPy array.

```
r = 0.025           # interest rate
T = 50              # end date
b = np.empty(T+1)   # an empty NumPy array, to store all b_t
b[0] = 10           # initial balance

for t in range(T):
    b[t+1] = (1 + r) * b[t]

plt.plot(b, label='bank balance')
plt.legend()
plt.show()
```

The statement b = np.empty(T+1) allocates storage in memory for T+1 (floating point) numbers.

These numbers are filled in by the for loop.

Allocating memory at the start is more efficient than using a Python list and append, since the latter must repeatedly ask for storage space from the operating system.

Notice that we added a legend to the plot—a feature you will be asked to use in the exercises.

3.6 Exercises

Now we turn to exercises. It is important that you complete them before continuing, since they present new concepts we will need.

3.6.1 Exercise 1

Your first task is to simulate and plot the correlated time series

$$x_{t+1} = \alpha\, x_t + \epsilon_{t+1} \quad \text{where} \quad x_0 = 0 \quad \text{and} \quad t = 0, \ldots, T$$

The sequence of shocks $\{\epsilon_t\}$ is assumed to be IID and standard normal.

In your solution, restrict your import statements to

```
import numpy as np
import matplotlib.pyplot as plt
```

Set $T = 200$ and $\alpha = 0.9$.

3.6.2 Exercise 2

Starting with your solution to exercise 2, plot three simulated time series, one for each of the cases $\alpha = 0$, $\alpha = 0.8$, and $\alpha = 0.98$.

Use a for loop to step through the α values.

If you can, add a legend, to help distinguish between the three time series.

Hints:

- If you call the plot() function multiple times before calling show(), all of the lines you produce will end up on the same figure.
- For the legend, note that the expression 'foo' + str(42) evaluates to 'foo42'.

3.6.3   Exercise 3

Similar to the previous exercises, plot the time series

$$x_{t+1} = \alpha \, |x_t| + \epsilon_{t+1} \quad \text{where} \quad x_0 = 0 \quad \text{and} \quad t = 0, \ldots, T$$

Use $T = 200$, $\alpha = 0.9$ and $\{\epsilon_t\}$ as before.
  Search online for a function that can be used to compute the absolute value $|x_t|$.

3.6.4   Exercise 4

One important aspect of essentially all programming languages is branching and
conditions.
  In Python, conditions are usually implemented with an if–else syntax.
  Here is an example that prints -1 for each negative number in an array and 1 for
each nonnegative number

```
numbers = [-9, 2.3, -11, 0]
```

```
for x in numbers:
    if x < 0:
        print(-1)
    else:
        print(1)
```

Now, write a new solution to Exercise 3 that does not use an existing function to
compute the absolute value.
  Replace this existing function with an if–else condition.

3.6.5   Exercise 5

Here is a harder exercise that takes some thought and planning.
  The task is to compute an approximation to π using Monte Carlo.
  Use no imports besides

```
import numpy as np
```

Your hints are as follows:

- If U is a bivariate uniform random variable on the unit square $(0, 1)^2$, then the
  probability that U lies in a subset B of $(0, 1)^2$ is equal to the area of B.

- If $U_1, \ldots, U_n$ are IID copies of U, then, as n gets large, the fraction that falls in B converges to the probability of landing in B.
- For a circle, $area = \pi * radius^2$.

3.7 Solutions

3.7.1 Exercise 1

Here is one solution.

```python
a = 0.9
T = 200
x = np.empty(T+1)
x[0] = 0

for t in range(T):
    x[t+1] = a * x[t] + np.random.randn()

plt.plot(x)
plt.show()
```

3.7.2 Exercise 2

```python
a_values = [0.0, 0.8, 0.98]
T = 200
x = np.empty(T+1)

for a in a_values:
    x[0] = 0
    for t in range(T):
        x[t+1] = a * x[t] + np.random.randn()
    plt.plot(x, label=f'$\\alpha = {a}$')

plt.legend()
plt.show()
```

3.7.3 Exercise 3

Here's one solution:

```
a = 0.9
T = 200
x = np.empty(T+1)
x[0] = 0

for t in range(T):
    x[t+1] = a * np.abs(x[t]) + np.random.randn()

plt.plot(x)
plt.show()
```

3.7.4 Exercise 4

Here is one way:

```
a = 0.9
T = 200
x = np.empty(T+1)
x[0] = 0

for t in range(T):
    if x[t] < 0:
        abs_x = - x[t]
    else:
        abs_x = x[t]
    x[t+1] = a * abs_x + np.random.randn()

plt.plot(x)
plt.show()
```

Here is a shorter way to write the same thing:

```
a = 0.9
T = 200
x = np.empty(T+1)
x[0] = 0

for t in range(T):
    abs_x = - x[t] if x[t] < 0 else x[t]
    x[t+1] = a * abs_x + np.random.randn()

plt.plot(x)
plt.show()
```

3.7.5 Exercise 5

Consider the circle of diameter 1 embedded in the unit square.

Let A be its area, and let $r = 1/2$ be its radius.

If we know π, then we can compute A via $A = \pi r^2$.

But here the point is to compute π, which we can do by $\pi = A/r^2$.

Summary: If we can estimate the area of a circle with diameter 1, then dividing by $r^2 = (1/2)^2 = 1/4$ gives an estimate of π.

We estimate the area by sampling bivariate uniforms and looking at the fraction that falls into the circle.

```python
n = 100000

count = 0
for i in range(n):
    u, v = np.random.uniform(), np.random.uniform()
    d = np.sqrt((u - 0.5)**2 + (v - 0.5)**2)
    if d < 0.5:
        count += 1

area_estimate = count / n

print(area_estimate * 4)   # dividing by radius**2
```

Chapter 4
Basic Python

Abstract Basic Python programming begins with an understanding of the fundamental variables, types, and operators used in the language, as well as how to respond to errors. Basic types introduced will include Numbers, Lists, Strings, Sets, and Dictionaries. Different operations can be applied to the different data types. Learn the rules of Python programming and perform simple tasks with variables, types, and operators through the exercises provided. Learning error handling concepts will help to capture and respond to exceptions efficiently.

Learning outcomes:

- Write your first line of code.
- Learn and apply the core Python variables, data types, and operators.
- Employ variables, data types, and operators in loops and logical conditions to perform simple tasks.
- Capture and respond efficiently to exceptions.
- Use sets to extract unique data from lists.

After a brief introduction in Chap. 3, let us do a recap of what we have learnt in this chapter. We will be deep diving into features in Python in the next chapter.

Python is an interpreted language with a very simple syntax. The first step in learning any new language is producing "Hello, World!".

You will have opened a new notebook in Jupyter Notebook:

T. T. Teoh, Z. Rong, *Artificial Intelligence with Python*,
Machine Learning: Foundations, Methodologies, and Applications,
https://doi.org/10.1007/978-981-16-8615-3_4

Note two things:

- The drop-down menu item that says `code`.
- The layout of the text entry box below that (which starts with `In [] :`).

This specifies that this block is to be used for code. You can also use it for structured text blocks like this one by selecting `markdown` from the menu list. This tutorial will not teach `markdown`, but you can learn more here.

The tutorial that follows is similar to Learn Python's, although here you will run the examples on your own computer.

4.1 Hello, World!

For now, let us start with "Hello, World!".

Type the below text into the code block in your notebook and hit `Ctrl-Enter` to execute the code:

print("Hello, World!)

```
print("Hello, World!")
```

```
Hello, World!
```

You just ran your first program. See how Jupyter performs code highlighting for you, changing the colors of the text depending on the nature of the syntax used. `print` is a protected term and gets highlighted in green.

This also demonstrates how useful Jupyter Notebook can be. You can treat this just like a document, saving the file and storing the outputs of your program in the notebook. You could even email this to someone else and, if they have Jupyter Notebook, they could run the notebook and see what you have done.

From here on out, we will simply move through the Python coding tutorial and learn syntax and methods for coding.

4.2 Indentation

Python uses indentation to indicate parts of the code that needs to be executed together. Both tabs and spaces (usually four per level) are supported, and my preference is for tabs. This is the subject of mass debate, but do not worry about it. Whatever you decide to do is fine, but **do not—under any circumstances—mix tab indentation with space indentation.**

In this exercise, you will assign a value to a variable, check to see if a comparison is true, and then—based on the result—print.

First, spot the little + symbol on the menu bar just after the `save` symbol. Click that and you will get a new box to type the following code into. When you are done, press `Ctrl-Enter` or the `Run` button.

```
x = 1
if x == 1:
    # Indented ... and notice how Jupyter Notebook⎵
↪automatically indented for you
    print("x is 1")
```

```
x is 1
```

Any non-protected text term can be a variable. Please take note of the naming convention for Python's variable. x could just have easily been rewritten as your name. Usually, it is good practice to name our variables as descriptively as possible. This allows us to read algorithms/code like text (i.e., the code describes itself). It helps other people to understand the code you have written as well.

- To assign a variable with a specific value, use =.
- To test whether a variable has a specific value, use the Boolean operators:

  - equal: ==
  - not equal: !=
  - greater than: >
  - less than: <

- You can also add helpful comments to your code with the # symbol. Any line starting with a # is not executed by the interpreter. Personally, I find it very useful to make detailed notes about my thinking since, often, when you come back to code later you cannot remember why you did, what you did, or what your code is even supposed to do. This is especially important in a team setting when different members are contributing to the same codebase.

4.3 Variables and Types

Python is not "statically-typed." This means you do not have to declare all your variables before you can use them. You can create new variables whenever you want. Python is also "object-oriented," which means that every variable is an object. That will become more important the more experience you get.

Let us go through the core types of variables:

4.3.1   Numbers

Python supports two data types for numbers: integers and floats. Integers are whole numbers (e.g., 7), while floats are fractional (e.g., 7.321). You can also convert integers to floats, and vice versa, but you need to be aware of the risks of doing so.

Follow along with the code:

```python
integer = 7
print(integer)
# notice that the class printed is currently int
print(type(integer))
```

```
7
<class 'int'>
```

```python
float_ = 7.0
print(float_)
# Or you could convert the integer you already have
myfloat = float(integer)
# Note how the term `float` is green. It's a protected term.
print(myfloat)
```

```
7.0
7.0
```

```python
# Now see what happens when you convert a float to an int
myint = int(7.3)
print(myint)
```

```
7
```

Note how you lost precision when you converted a float to an int? Always be careful, since that could be the difference between a door that fits its frame and the one that is far too small.

4.3.2   Strings

Strings are the Python term for text. You can define these in either single or double quotes. I will be using double quotes (since you often use a single quote inside text phrases).

Try these examples:

```
mystring = "Hello, World!"
print(mystring)
# and demonstrating how to use an apostrophe in a string
mystring = "Let's talk about apostrophes..."
print(mystring)
```

```
Hello, World!
Let's talk about apostrophes...
```

You can also apply simple operators to your variables or assign multiple variables simultaneously.

```
one = 1
two = 2
three = one + two
print(three)

hello = "Hello,"
world = "World!"
helloworld = hello + " " + world
print(helloworld)

a, b = 3, 4
print(a, b)
```

```
3
Hello, World!
3 4
```

Note, though, that mixing variable data types causes problems.

```
print(one + two + hello)
```

```
---------------------------------------------------------------
↪------------
TypeError                                  Traceback (most␣
↪recent call last)
<ipython-input-8-a2961e4891f2> in <module>
----> 1 print(one + two + hello)

TypeError: unsupported operand type(s) for +: 'int' and 'str'
```

Python will throw an error when you make a mistake like this and the error will give you as much detail as it can about what just happened. This is extremely useful when you are attempting to "debug" your code.

In this case, you are told: `TypeError: unsupported operand type(s) for +: 'int' and 'str'`.

And the context should make it clear that you tried to combine two integer variables with a string.

You can also combine strings with placeholders for variables:

- Add variables to a string with `format`, e.g., `"Variable {}".format(x)` will replace the `{}` in the string with the value in the variable x.
- Floating point numbers can get out of hand (imagine including a number with 30 decimal places in a string), and you can format this with `"Variable {:10.4f}".format(x)`, where `10` is the amount of space allocated to the float (useful if you need to align a column of numbers; you can also leave this out to include all significant digits) and the `.4f` is the number of decimals after the point. Vary these as you need.

```
variable = 1/3 * 100
print("Unformated variable: {}%".format(variable))
print("Formatted variable: {:.3f}%".format(variable))
print("Space formatted variable: {:10.1f}%".format(variable))
```

```
Unformated variable: 33.33333333333333%
Formatted variable: 33.333%
Space formatted variable:         33.3%
```

4.3.3 Lists

`Lists` are an ordered list of any type of variable. You can combine as many variables as you like, and they could even be of multiple data types. Ordinarily, unless you have a specific reason to do so, lists will contain variables of one type.

You can also iterate over a list (use each item in a list in sequence).

A list is placed between square brackets: `[]`.

```
mylist = []
mylist.append(1)
mylist.append(2)
mylist.append(3)
# Each item in a list can be addressed directly.
# The first address in a Python list starts at 0
print(mylist[0])
# The last item in a Python list can be addressed as -1.
# This is helpful when you don't know how long a list is
↪likely to be.
print(mylist[-1])
# You can also select subsets of the data in a list like this
print(mylist[1:3])

# You can also loop through a list using a `for` statement.
# Note that `x` is a new variable which takes on the value of
↪each item in the list in order.
```

(continues on next page)

(continued from previous page)

```
for x in mylist:
    print(x)
```

```
1
3
[2, 3]
1
2
3
```

If you try to access an item in a list that is not there, you will get an error.

```
print(mylist[10])
```

```
---------------------------------------------------------------
↪------------
IndexError                              Traceback (most↪
↪recent call last)
<ipython-input-11-46c3ae90a572> in <module>
----> 1 print(mylist[10])

IndexError: list index out of range
```

Let us put this together with a slightly more complex example. But first, some new syntax:

- Check what type of variable you have with isinstance, e.g., isinstance(x, float) will be True if x is a float.
- You have already seen for, but you can get the loop count by wrapping your list in the term enumerate, e.g., for count, x in enumerate(mylist) will give you a count for each item in the list.
- Sorting a list into numerical or alphabetical order can be done with sort.
- Getting the number of items in a list is as simple as asking len(list).
- If you want to count the number of times a particular variable occurs in a list, use list.count(x) (where x is the variable you are interested in).

Try this for yourself.

```
# Let's imagine we have a list of unordered names that somehow↪
↪got some random numbers included.
# For this exercise, we want to print the alphabetised list of↪
↪names without the numbers.
# This is not the best way of doing the exercise, but it will↪
↪illustrate a whole bunch of techniques.
names = ["John", 3234, 2342, 3323, "Eric", 234, "Jessica",↪
↪734978234, "Lois", 2384]
print("Number of names in list: {}".format(len(names)))
# First, let's get rid of all the weird integers.
new_names = []
```

(continues on next page)

(continued from previous page)

```python
for n in names:
    if isinstance(n, str):
        # Checking if n is a string
        # And note how we're now indented twice into this new
↪component
            new_names.append(n)
# We should now have only names in the new list. Let's sort
↪them.
new_names.sort()
print("Cleaned-up number of names in list: {}".format(len(new_
↪names)))
# Lastly, let's print them.
for i, n in enumerate(new_names):
    # Using both i and n in a formated string
    # Adding 1 to i because lists start at 0
    print("{}. {}".format(i+1, n))
```

```
Number of names in list: 10
Cleaned-up number of names in list: 4
1. Eric
2. Jessica
3. John
4. Lois
```

4.3.4 Dictionaries

Dictionaries are one of the most useful and versatile data types in Python. They are similar to arrays but consist of key:value pairs. Each value stored in a dictionary is accessed by its key, and the value can be any sort of object (string, number, list, etc.).

This allows you to create structured records. Dictionaries are placed within { }.

```python
phonebook = {}
phonebook["John"] = {"Phone": "012 794 794",
                     "Email": "john@email.com"}
phonebook["Jill"] = {"Phone": "012 345 345",
                     "Email": "jill@email.com"}
phonebook["Joss"] = {"Phone": "012 321 321",
                     "Email": "joss@email.com"}
print(phonebook)
```

```
{'John': {'Phone': '012 794 794', 'Email': 'john@email.com'},
 ↪'Jill': {'Phone': '012 345 345', 'Email': 'jill@email.com'},
 ↪'Joss': {'Phone': '012 321 321', 'Email': 'joss@email.com'}}
```

Note that you can nest dictionaries and lists. The above shows you how you can add values to an existing dictionary or create dictionaries with values.

You can iterate over a dictionary just like a list, using the dot term `.items()`. In Python 3, the dictionary maintains the order in which data were added, but older versions of Python do not.

```
for name, record in phonebook.items():
    print("{}'s phone number is {}, and their email is {}".
↪format(name, record["Phone"], record["Email"]))
```

```
John's phone number is 012 794 794, and their email is␣
↪john@email.com
Jill's phone number is 012 345 345, and their email is␣
↪jill@email.com
Joss's phone number is 012 321 321, and their email is␣
↪joss@email.com
```

You add new records as shown above, and you remove records with `del` or `pop`. They each have a different effect.

```
# First `del`
del phonebook["John"]
for name, record in phonebook.items():
    print("{}'s phone number is {}, and their email is {}".
↪format(name, record["Phone"], record["Email"]))

# Pop returns the record, and deletes it
jill_record = phonebook.pop("Jill")
print(jill_record)
for name, record in phonebook.items():
    # You can see that only Joss is still left in the system
    print("{}'s phone number is {}, and their email is {}".
↪format(name, record["Phone"], record["Email"]))

# If you try and delete a record that isn't in the dictionary,␣
↪you get an error
del phonebook["John"]
```

```
Jill's phone number is 012 345 345, and their email is␣
↪jill@email.com
Joss's phone number is 012 321 321, and their email is␣
↪joss@email.com
{'Phone': '012 345 345', 'Email': 'jill@email.com'}
Joss's phone number is 012 321 321, and their email is␣
↪joss@email.com
```

```
---------------------------------------------------------------
↪------------
KeyError                                    Traceback (most␣
↪recent call last)
<ipython-input-15-24fe1d3ad744> in <module>
     12
     13 # If you try and delete a record that isn't in the␣
↪dictionary, you get an error
```

(continues on next page)

(continued from previous page)

```
---> 14 del phonebook["John"]

KeyError: 'John'
```

One thing to get into the habit of doing is to test variables before assuming they have characteristics you are looking for. You can test a dictionary to see if it has a record and return some default answer if it does not have it.

You do this with the `.get("key", default)` term. Default can be anything, including another variable, or simply `True` or `False`. If you leave default blank (i.e., `.get("key")`), then the result will automatically be `False` if there is no record.

```
# False and True are special terms in Python that allow you to
↪set tests
jill_record = phonebook.get("Jill", False)
if jill_record: # i.e. if you got a record in the previous step
    print("Jill's phone number is {}, and their email is {}".
↪format(jill_record["Phone"], jill_record["Email"]))
else: # the alternative, if `if` returns False
    print("No record found.")
```

```
No record found.
```

4.4   Basic Operators

Operators are the various algebraic symbols (such as +, -, *, /, %, etc.). Once 'you have learned the syntax, programming is mostly mathematics.

4.4.1   Arithmetic Operators

As you would expect, you can use the various mathematical operators with numbers (both integers and floats).

```
number = 1 +2 * 3 / 4.0
# Try to predict what the answer will be ... does Python
↪follow order operations hierarchy?
print(number)

# The modulo (%) returns the integer remainder of a division
remainder = 11 % 3
print(remainder)
```

(continues on next page)

(continued from previous page)

```
# Two multiplications is equivalent to a power operation
squared = 7 ** 2
print(squared)
cubed = 2 ** 3
print(cubed)
```

```
2.5
2
49
8
```

4.4.2 List Operators

```
even_numbers = [2, 4, 6, 8]
# One of my first teachers in school said, "People are odd.␣
↪Numbers are uneven."
# He also said, "Cecil John Rhodes always ate piles of␣
↪unshelled peanuts in parliament in Cape Town."
# "You'd know he'd been in parliament by the huge pile of␣
↪shells on the floor. He also never wore socks."
# "You'll never forget this." And I didn't. I have no idea if␣
↪it's true.
uneven_numbers = [1, 3, 5, 7]
all_numbers = uneven_numbers + even_numbers
# What do you think will happen?
print(all_numbers)

# You can also repeat sequences of lists
print([1, 2 , 3] * 3)
```

```
[1, 3, 5, 7, 2, 4, 6, 8]
[1, 2, 3, 1, 2, 3, 1, 2, 3]
```

We can put this together into a small project.

```
x = object() # A generic Python object
y = object()

# Change this code to ensure that x_list and y_list each have␣
↪10 repeating objects
# and concat_list is the concatenation of x_list and y_list
x_list = [x]
y_list = [y]
concat_list = []

print("x_list contains {} objects".format(len(x_list)))
```

(continues on next page)

(continued from previous page)

```
print("y_list contains {} objects".format(len(y_list)))
print("big_list contains {} objects".format(len(concat_list)))

# Test your lists
if x_list.count(x) == 10 and y_list.count(y) == 10:
    print("Almost there...")
if concat_list.count(x) == 10 and concat_list.count(y) == 10:
    print("Great!")
```

```
x_list contains 1 objects
y_list contains 1 objects
big_list contains 0 objects
```

4.4.3   String Operators

You can do a surprising amount with operators on strings.

```
# You've already seen arithmetic concatenations of strings
helloworld = "Hello," + " " + "World!"
print(helloworld)

# You can also multiply strings to form a repeating sequence
manyhellos = "Hello " * 10
print(manyhellos)

# But don't get carried away. Not everything will work.
nohellos = "Hello " / 10
print(nohellos)
```

```
Hello, World!
Hello Hello Hello Hello Hello Hello Hello
```

```
----------------------------------------------------------------
↪------------
TypeError                                       Traceback (most↲
↪recent call last)
<ipython-input-20-b66f17ae47c7> in <module>
      8
      9 # But don't get carried away. Not everything will work.
---> 10 nohellos = "Hello " / 10
     11 print(nohellos)

TypeError: unsupported operand type(s) for /: 'str' and 'int'
```

Something to keep in mind is that strings are lists of characters. This means you can perform a number of list operations on strings. Additionally, there are a few more new operations that you can perform on strings compared to lists.

- Get the index for the first occurrence of a specific letter with `string.index("l")`, where l is the letter 'you are looking for.
- As in lists, count the number of occurrences of a specific letter with `string.count("l")`.
- Get slices of strings with `string[start:end]`, e.g., `string[3:7]`. If 'you are unsure of the end of a string, remember you can use negative numbers to count from the end, e.g., `string[:-3]` to get a slice from the first character to the third from the end.
- You can also "step" through a string with `string[start:stop:step]`, e.g., `string[2:6:2]`, which will skip a character between the characters 2 and 5 (i.e., 6 is the boundary).
- You can use a negative "step" to reverse the order of the characters, e.g., `string[::-1]`.
- You can convert strings to upper- or lower-case with `string.upper()` and `string.lower()`.
- Test whether a string starts or ends with a substring with:

  - `string.startswith(substring)` which returns True or False
  - `string.endswith(substring)` which returns True or False

- Use in to test whether a string contains a substring, so `substring in string` will return True or False.
- You can split a string into a genuine list with `.split(s)`, where s is the specific character to use for splitting, e.g., `s = ","` or `s = " "`. You can see how this might be useful to split up text which contains numeric data.

```
a_string = "Hello, World!"
print("String length: {}".format(len(a_string)))
# You can get an index of the first occurrence of a specific⌴
↪letter
# Remember that Python lists are based at 0; the first letter⌴
↪is index 0
# Also note the use of single quotes inside the double quotes
print("Index for first 'o': {}".format(a_string.index("o")))
print("Count of 'o': {}".format(a_string.count("o")))
print("Slicing between second and fifth characters: {}".
↪format(a_string[2:6]))
print("Skipping between 3rd and 2nd-from-last characters: {}".
↪format(a_string[3:-2:2]))
print("Reverse text: {}".format(a_string[::-1]))
print("Starts with 'Hello': {}".format(a_string.startswith(
↪"Hello")))
print("Ends with 'Hello': {}".format(a_string.endswith("Hello
↪")))
print("Contains 'Goodbye': {}".format("Goodby" in a_string))
print("Split the string: {}".format(a_string.split(" ")))
```

```
String length: 13
Index for first 'o': 4
Count of 'o': 2
Slicing between second and fifth characters: llo,
Skipping between 3rd and 2nd-from-last characters: l,Wr
Reverse text: !dlroW ,olleH
Starts with 'Hello': True
Ends with 'Hello': False
Contains 'Goodbye': False
Split the string: ['Hello,', 'World!']
```

4.5   Logical Conditions

In the section on *Indentation*, you were introduced to the if statement and the set
of boolean operators that allow you to test different variables against each other.

To that list of Boolean operators are added a new set of comparisons: and, or,
and in.

```
# Simple boolean tests
x = 2
print(x == 2)
print(x == 3)
print(x < 3)

# Using `and`
name = "John"
print(name == "John" and x == 2)

# Using `or`
print(name == "John" or name == "Jill")

# Using `in` on lists
print(name in ["John", "Jill", "Jess"])
```

```
True
False
True
True
True
True
```

These can be used to create nuanced comparisons using if. You can use a series
of comparisons with if, elif, and else.

Remember that code must be indented correctly or you will get unexpected
behavior.

```
# Unexpected results
x = 2
if x > 2:
    print("Testing x")
print("x > 2")
# Formated correctly
if x == 2:
    print("x == 2")
```

```
x > 2
x == 2
```

```
# Demonstrating more complex if tests
x = 2
y = 10
if x > 2:
    print("x > 2")
elif x == 2 and y > 50:
    print("x == 2 and y > 50")
elif x < 10 or y > 50:
    # But, remember, you don't know WHICH condition was True
    print("x < 10 or y > 50")
else:
    print("Nothing worked.")
```

```
x < 10 or y > 50
```

Two special cases are not and is.

- not is used to get the opposite of a particular Boolean test, e.g., not(False) returns True.
- is would seem, superficially, to be similar to ==, but it tests for whether the actual objects are the same, not whether the values which the objects reflect are equal.

A quick demonstration:

```
# Using `not`
name_list1 = ["John", "Jill"]
name_list2 = ["John", "Jill"]
print(not(name_list1 == name_list2))

# Using `is`
name2 = "John"
print(name_list1 == name_list2)
print(name_list1 is name_list2)
```

```
False
True
False
```

4.6 Loops

Loops iterate over a given sequence, and—here—it is critical to ensure your indentation is correct or 'you will get unexpected results for what is considered inside or outside the loop.

- For loops, for, which loop through a list. There is also some new syntax to use in for loops:
  - In *Lists* you saw enumerate, which allows you to count the loop number.
  - Range creates a list of integers to loop, range(start, stop) creates a list of integers between start and stop, or range(num) creates a zero-based list up to num, or range(start, stop, step) steps through a list in increments of step.
- While loops, while, which execute while a particular condition is True. And some new syntax for while is:
  - while is a conditional statement (it requires a test to return True), which means we can use else in a while loop (but not for)

```
# For loops

for i, x in enumerate(range(2, 8, 2)):
    print("{}. Range {}".format(i+1, x))

# While loops
count = 0
while count < 5:
    print(count)
    count += 1 # A shorthand for count = count + 1
else:
    print("End of while loop reached")
```

```
1. Range 2
2. Range 4
3. Range 6
0
1
2
3
4
End of while loop reached
```

Pay close attention to the indentation in that while loop. What would happen if count += 1 were outside the loop?

What happens if you need to exit loops early or miss a step?

- break exits a while or for loop immediately.
- continue skips the current loop and returns to the loop conditional.

```
# Break and while conditional
print("Break and while conditional")
count = 0
while True:
    # You may think this would run forever, but ...
    print(count)
    count += 1
    if count >= 5:
        break

# Continue
print("Continue")
for x in range(8):
    # Check if x is uneven
    if (x+1) % 2 == 0:
        continue
    print(x)
```

```
Break and while conditional
0
1
2
3
4
Continue
0
2
4
6
```

4.7 List Comprehensions

One of the common tasks in coding is to go through a list of items, edit or apply some form of algorithm, and return a new list.

Writing long stretches of code to accomplish this is tedious and time-consuming. List comprehensions are an efficient and concise way of achieving exactly that.

As an example, imagine we have a sentence where we want to count the length of each word but skip all the "the"s:

```
sentence = "for the song and the sword are birthrights sold to↵
↪an usurer, but I am the last lone highwayman and I am the↵
↪last adventurer"
words = sentence.split()
word_lengths = []
for word in words:
    if word != "the":
        word_lengths.append(len(word))
```

(continues on next page)

(continued from previous page)

```
print(word_lengths)

# The exact same thing can be achieved with a list␣
↪comprehension
word_lengths = [len(word) for word in sentence.split(" ") if␣
↪word != "the"]
print(word_lengths)
```

```
[3, 4, 3, 5, 3, 11, 4, 2, 2, 7, 3, 1, 2, 4, 4, 10, 3, 1, 2, 4,␣
↪10]
[3, 4, 3, 5, 3, 11, 4, 2, 2, 7, 3, 1, 2, 4, 4, 10, 3, 1, 2, 4,␣
↪10]
```

4.8 Exception Handling

For the rest of this section of the tutorial, 'we are going to focus on some more advanced syntax and methodology.

In [Python basics: Strings](02—Python basics.ipynb#Strings), you say how the instruction to concatenate a string with an integer using the + operator resulted in an error:

```
print(1 + "hello")
```

```
--------------------------------------------------------------
↪------------
TypeError                                        Traceback (most␣
↪recent call last)
<ipython-input-29-778236ccec55> in <module>
----> 1 print(1 + "hello")

TypeError: unsupported operand type(s) for +: 'int' and 'str'
```

In Python, this is known as an exception. The particular exception here is a TypeError. Getting exceptions is critical to coding, because it permits you to fix syntax errors, catch glitches where you pass the wrong variables, or your code behaves in unexpected ways.

However, once your code goes into production, errors that stop your program entirely are frustrating for the user. More often than not, there is no way to exclude errors and sometimes the only way to find something out is to try it and see if the function causes an error.

Many of these errors are entirely expected. For example, if you need a user to enter an integer, you want to prevent them typing in text, or—in this era of mass hacking—you want to prevent a user trying to include their own code in a text field.

When this happens what you want is to way to try to execute your code and then catch any expected exceptions safely.

- Test and catch exceptions with `try` and `except`.
- Catch specific errors, rather than all errors, since you still need to know about anything unexpected; otherwise, you can spend hours trying to find a mistake which is being deliberately ignored by your program.
- Chain exceptions with, e.g., `except (IndexError, TypeError):`. Here is a link to all the common exceptions.

```
# An `IndexError` is thrown when you try to address an index
→in a list that does not exist
# In this example, let's catch that error and do something else

def print_list(l):
    """
    For a given list `l`, of unknown length, try to print out
→the first
    10 items in the list.

    If the list is shorter than 10, fill in the remaining
→items with `0`.
    """
    for i in range(10):
        try:
            print(l[i])
        except IndexError:
            print(0)

print_list([1,2,3,4,5,6,7])
```

```
1
2
3
4
5
6
7
0
0
0
```

You can also deliberately trigger an exception with `raise`. To go further and write your own types of exceptions, consider this explanation.

```
def print_zero(zero):
    if zero != 0:
        raise ValueError("Not Zero!")
    print(zero)

print_zero(10)
```

```
-----------------------------------------------------------------
↪------------
ValueError                                       Traceback (most␣
↪recent call last)
<ipython-input-1-a5b85de8b30c> in <module>
      4        print(zero)
      5
----> 6 print_zero(10)

<ipython-input-1-a5b85de8b30c> in print_zero(zero)
      1 def print_zero(zero):
      2        if zero != 0:
----> 3            raise ValueError("Not Zero!")
      4        print(zero)
      5

ValueError: Not Zero!
```

4.8.1   Sets

Sets are lists with no duplicate entries. You could probably write a sorting algorithm, or dictionary, to achieve the same end, but sets are faster and more flexible.

```
# Extract all unique terms in this sentence

print(set("the rain is wet and wet is the rain".split()))
```

```
{'rain', 'and', 'is', 'the', 'wet'}
```

- Create a unique set of terms with set.
- To get members of a set common to both of two sets, use set1.intersection(set2).
- Get the unique members of each of one set and another, use set1.symmetric_difference(set2).
- To get the unique members from the asking set (i.e., the one calling the dot function), use set1.difference(set2).
- To get all the members of each of two lists, use set1.union(set2).

```
set_one = set(["Alice", "Carol", "Dan", "Eve", "Heidi"])
set_two = set(["Bob", "Dan", "Eve", "Grace", "Heidi"])

# Intersection
print("Set One intersection: {}".format(set_one.
↪intersection(set_two)))
print("Set Two intersection: {}".format(set_two.
↪intersection(set_one)))
```

(continues on next page)

(continued from previous page)

```
# Symmetric difference
print("Set One symmetric difference: {}".format(set_one.
↪symmetric_difference(set_two)))
print("Set Two symmetric difference: {}".format(set_two.
↪symmetric_difference(set_one)))

# Difference
print("Set One difference: {}".format(set_one.difference(set_
↪two)))
print("Set Two difference: {}".format(set_two.difference(set_
↪one)))

# Union
print("Set One union: {}".format(set_one.union(set_two)))
print("Set Two union: {}".format(set_two.union(set_one)))
```

```
Set One intersection: {'Dan', 'Heidi', 'Eve'}
Set Two intersection: {'Dan', 'Heidi', 'Eve'}
Set One symmetric difference: {'Carol', 'Grace', 'Alice', 'Bob
↪'}
Set Two symmetric difference: {'Grace', 'Carol', 'Alice', 'Bob
↪'}
Set One difference: {'Alice', 'Carol'}
Set Two difference: {'Bob', 'Grace'}
Set One union: {'Carol', 'Grace', 'Heidi', 'Dan', 'Alice', 'Eve
↪', 'Bob'}
Set Two union: {'Grace', 'Carol', 'Heidi', 'Dan', 'Alice', 'Eve
↪', 'Bob'}
```

Chapter 5
Intermediate Python

Abstract Codes covered in Basic Python were generally limited to short snippets. Intermediate Python will cover concepts that will allow codes to be reused such as developing functions, classes, objects, modules, and packages. Functions are callable modules of code that are written to take in parameters to perform a task and return a value. Decorators and closures can be applied to modify function behaviors. Objects encapsulate both variables and functions into a single entity which are defined in classes. A module in Python is a set of classes or functions that encapsulate a single, and related, set of tasks. Packages are a set of modules collected together into a single focused unit.

Learning outcomes:

- Develop and use reusable code by encapsulating tasks in functions.
- Package functions into flexible and extensible classes.
- Apply closures and decorators to functions to modify function behavior.

The code from the previous chapter (Python syntax (basic)) was limited to short snippets. Solving more complex problems means more complex code stretching over hundreds, to thousands, of lines, and—if you want to reuse that code—it is not convenient to copy and paste it multiple times. Worse, any error is magnified, and any changes become tedious to manage.

Imagine that you have a piece of code that you have written, and it has been used in many different files. It would be troublesome to modify every single files when we have to change the function. If the code used is critical for the application, it would be disastrous to miss out changes to the code. It would be far better to write a discrete module to contain that task, get it absolutely perfect, and call it whenever you want the same problem solved or task executed.

In software, this process of packaging up discrete functions into their own modular code is called "abstraction." A complete software system consists of a number of discrete modules all interacting to produce an integrated experience.

In Python, these modules are called `functions` and a complete suite of functions grouped around a set of related tasks is called a `library` or `module`. Libraries permit you to inherit a wide variety of powerful software solutions developed and maintained by other people.

T. T. Teoh, Z. Rong, *Artificial Intelligence with Python,*
Machine Learning: Foundations, Methodologies, and Applications,
https://doi.org/10.1007/978-981-16-8615-3_5

Python is open source, which means that its source code is released under a license which permits anyone to study, change, and distribute the software to anyone and for any purpose. Many of the most popular Python libraries are also open source. There are thousands of shared libraries for you to use and—maybe, when you feel confident enough—to contribute to with your own code.

5.1   Functions

Functions are callable modules of code, some with parameters or arguments (variables you can pass to the function), which performs a task and may return a value. They are a convenient way to package code into discrete blocks, making your overall program more readable, reusable, and saving time.

You can also easily share your functions with others, saving them time as well.

- You structure a function using `def`, like so: `def function_name(parameters): code return response`.
- `return` is optional but allows you to return the results of any task performed by the function to the point where the function was called.
- To test whether an object is a function (i.e., callable), use `callable`, e.g., `callable(function)` will return 1.

```python
# A simple function with no arguments
def say_hello():
    print("Hello, World!")

# Calling it is as simple as this
say_hello()

# And you can test that it's a callable
print(callable(say_hello))

# () <- means calling a function and thus callable checks if
↪something is callable
```

```
Hello, World!
True
```

An argument can be any variable, such as integers, strings, lists, dictionaries, or even other functions. This is where you start realizing the importance of leaving comments and explanations in your code because you need to ensure that anyone using a function knows what variables the function expects and in what order.

Functions can also perform calculations and return these to whatever called them.

```python
# A function with two string arguments
def say_hello_to_user(username, greeting):
```

(continues on next page)

(continued from previous page)

```
    # Returns a greeting to a username
    print("Hello, {}! I hope you have a great {}.".
↪format(username, greeting))

# Call it
say_hello_to_user("Jill", "day")

# Perform a calculation and return it
def sum_two_numbers(x, y):
    # Returns the sum of x + y
    return x + y

sum_two_numbers(5, 10)
```

```
Hello, Jill! I hope you have a great day.
```

```
15
```

You can see that swapping `username` and `greeting` in the `say_hello_to_user` function would be confusing, but swapping the numbers in `sum_two_numbers` would not cause a problem.

Not only can you call functions from functions, but you can also create variables that are functions or the result of functions.

```
def number_powered(number, exponent):
    # Returns number to the power of exponent
    return number ** exponent

# Jupyter keeps functions available that were called in other↵
↪cells
# This means `sum_two_numbers` is still available
def sum_and_power(number1, number2, exponent):
    # Returns two numbers summed, and then to an exponent
    summed = sum_two_numbers(number1, number2)
    return number_powered(summed, exponent)

# Call `sum_and_power`
print(sum_and_power(2, 3, 4))
```

```
625
```

With careful naming and plenty of commentary, you can see how you can make your code extremely readable and self-explanatory.

A better way of writing comments in functions is called `docstrings`.

- Docstrings are written as structured text between three sets of inverted commas, e.g., `""" This is a docstring """`.
- You can access a function's docstring by calling `function.__doc__`.

```python
def docstring_example():
    """
    An example function which returns `True`.
    """
    return True

# Printing the docstring
print(docstring_example.__doc__)

# Calling it
print(docstring_example())
```

```
    An example function which returns `True`.

True
```

5.2 Classes and Objects

A complete Python object is an encapsulation of both variables and functions into a single entity. Objects get their variables and functions from `classes`.

Classes are where most of the action happens in Python and coding consists, largely, of producing and using classes to perform tasks.

A very basic class would look like this:

```python
class myClass:
    """
    A demonstration class.
    """
    my_variable = "Look, a variable!"

    def my_function(self):
        """
        A demonstration class function.
        """
        return "I'm a class function!"

# You call a class by creating a new class object
new_class = myClass()

# You can access class variables or functions with a dotted
# ↪call, as follows
print(new_class.my_variable)
print(new_class.my_function())

# Access the class docstrings
print(myClass.__doc__)
print(myClass.my_function.__doc__)
```

```
Look, a variable!
I'm a class function!

    A demonstration class.


        A demonstration class function.
```

Let us unpack the new syntax.

- You instantiate a class by calling it as class(). If you only called class, without the brackets, you would gain access to the object itself. This is useful as you can pass classes around as you would for variables.
- All variables and functions of a class are reached via the dotted call, .function() or .variable. You can even add new functions and variables to a class you created. Remember, though, these will not exist in new classes you create since you have not changed the underlying code.
- Functions within a class require a base argument that, by convention, is called self. There is a complex explanation as to why self is needed, but— briefly—think of it as the instance of the object itself. So, inside the class, self.function is the way the class calls its component functions.
- You can also access the docstrings as you would before.

```
# Add a new variable to a class instance
new_class1 = myClass()
new_class1.my_variable2 = "Hi, Bob!"
print(new_class1.my_variable2, new_class1.my_variable)

# But, trying to access my_variable2 in new_class causes an
↪error
print(new_class.my_variable2)
```

```
Hi, Bob! Look, a variable!
```

```
----------------------------------------------------------------
↪------------
AttributeError                                  Traceback (most
↪recent call last)
<ipython-input-6-544fecdbf014> in <module>()
      5
      6 # But, trying to access my_variable2 in new_class
↪causes an error
----> 7 print(new_class.my_variable2)

AttributeError: 'myClass' object has no attribute 'my_variable2
↪'
```

Classes can initialize themselves with a set of available variables. This makes the self referencing more explicit and also permits you to pass arguments to your class to set the initial values.

- Initialize a class with the special function def __init__(self).
- Pass arguments to your functions with __init__(self, arguments).
- We can also differentiate between arguments and keyword arguments:
  - **arguments**: these are passed in the usual way, as a single term, e.g., my_function(argument).
  - **keyword arguments**: these are passed the way you would think of a dictionary, e.g., my_function(keyword_argument = value). This is also a way to initialize an argument with a default. If you leave out the argument when it has a default, it will apply without the function failing.
  - Functions often need to have numerous arguments and keyword arguments passed to them, and this can get messy. You can also think of a list of arguments like a list and a list of keyword arguments like a dictionary. A tidier way to deal with this is to reference your arguments and keyword arguments like this, my_function(*args, **kwargs), where *args will be available to the function as an ordered list and **kwargs as a dictionary.

```
# A demonstration of all these new concepts

class demoClass:
    """
    A demonstration class with an __init__ function, and a
    ↪function that takes args and kwargs.
    """

    def __init__(self, argument = None):
        """
        A function that is called automatically when the
        ↪demoClass is initialised.
        """
        self.demo_variable = "Hello, World!"
        self.initial_variable = argument

    def demo_class(self, *args, **kwargs):
        """
        A demo class that loops through any args and kwargs
        ↪provided and prints them.
        """
        for i, a in enumerate(args):
            print("Arg {}: {}".format(i+1, a))
        for k, v in kwargs.items():
            print("{} - {}".format(k, v))
        if kwargs.get(self.initial_variable):
            print(self.demo_variable)
```

(continues on next page)

(continued from previous page)

```
        return True

demo1 = demoClass()
demo2 = demoClass("Bob")

# What was initialised in each demo object?
print(demo1.demo_variable, demo1.initial_variable)
print(demo2.demo_variable, demo2.initial_variable)

# A demo of passing arguments and keyword arguments
args = ["Alice", "Bob", "Carol", "Dave"]
kwargs = {"Alice": "Engineer",
          "Bob": "Consultant",
          "Carol": "Lawyer",
          "Dave": "Doctor"
         }
demo2.demo_class(*args, **kwargs)
```

```
Hello, World! None
Hello, World! Bob
Arg 1: Alice
Arg 2: Bob
Arg 3: Carol
Arg 4: Dave
Alice - Engineer
Bob - Consultant
Carol - Lawyer
Dave - Doctor
Hello, World!
```

```
True
```

Using *args and **kwargs in your function calls while you are developing makes it easier to change your code without having to go back through every line of code that calls your function and bug-fix when you change the order or number of arguments you are calling.

This reduces errors, improves readability, and makes for a more enjoyable and efficient coding experience.

At this stage, you have learned the fundamental syntax, as well as how to create modular code. Now we need to make our code reusable and shareable.

5.3 Modules and Packages

A module in Python is a set of classes or functions that encapsulate a single, and related, set of tasks. Packages are a set of modules collected together into a single focused unit. This can also be called a library.

Creating a module is as simple as saving your class code in a file with the `.py` extension (much as a text file ends with `.txt`).

5.3.1 Writing Modules

A set of modules in a library have a specific set of requirements. Imagine we wish to develop a ping pong game. We can place the game logic in one module and the functionality for drawing the game in another. That leads to a folder with the following file structure:

```
pingpong/
pingpong/game.py
pingpong/draw.py
```

Within each file will be a set of functions. Assume that, within `draw.py`, there is a function called `draw_game`. If you wanted to import the `draw_game` function into the `game.py` file, the convention is as follows:

```
import draw
```

This will import everything in the `draw.py` file. After that, you access functions from the file by making calls to, for example, `draw.draw_game`.

Or, you can access each function directly and only import what you need (since some files can be extremely large and you do not necessarily wish to import everything):

```
from draw import draw_game
```

You are not always going to want to run programs from an interpreter (like Jupyter Notebook). When you run a program directly from the command-line, you need a special function called `main`, which is then executed as follows:

```
if __name__ == '__main__':
    main()
```

Putting that together, the syntax for calling `game.py` from the command-line would be:

- Python functions and classes can be saved for reuse into files with the extension `.py`.
- You can import the functions from those files using either `import filename` (without the `.py` extension) or specific functions or classes from that file with `from filename import class, function1, function2`.
- You may notice that, after you run your program, Python automatically creates a file with the same name, but with `.pyc` as an extension. This is a compiled version of the file and happens automatically.

- If you intend to run a file from the command-line, you must insert a `main` function and call it as follows: `if __name__ == '__main__': main()`.
- If a module has a large number of functions you intend to use throughout your own code, then you can specify a custom name for use. For example, a module we will learn about in the next section is called `pandas`. The convention is to import it as `import pandas as pd`. Now you would access the functions in `pandas` using the dot notation of `pd.function`.
- You can also import modules based on logical conditions. If you import these options under the same name, your code is not affected by logical outcomes.

Putting all of this together in a pseudocode example (i.e., this code does not work, so do not try executing it):

```
# game.py
# Import the draw module
visual_mode = True
if visual_mode:
    # in visual mode, we draw using graphics
    import draw_visual as draw
else:
    # In textual mode, we print out text
    import draw_textual as draw

def main():
    result = play_game()
    # this can either be visual or textual depending on visual_
↪mode
    draw.draw_game(result)
```

```
-------------------------------------------------------------
↪------------
ModuleNotFoundError                          Traceback (most␣
↪recent call last)
<ipython-input-15-caaebde59de2> in <module>
      4 if visual_mode:
      5     # in visual mode, we draw using graphics
----> 6     import draw_visual as draw
      7 else:
      8     # In textual mode, we print out text

ModuleNotFoundError: No module named 'draw_visual'
```

Using the following pseudocode, the program will break. Note, though, that this shows how "safe" it is to experiment with code snippets in Jupyter Notebook. There is no harm done.

5.4   Built-in Modules

There are a vast range of built-in modules. Jupyter Notebook comes with an even larger list of third-party modules you can explore.

- After you have imported a module, `dir(module)` lets you see a list of all the functions implemented in that library.
- You can also read the help from the module docstrings with `help(module)`.

Let us explore a module you will be using and learning about in future sessions of this course, `pandas`.

We will print the top 1000 characters of `pandas` docstring.

```
import pandas as pd

help(pd)
```

```
Help on package pandas:

NAME
    pandas

DESCRIPTION
    pandas - a powerful data analysis and manipulation library␣
↪for Python

    ␣
↪============================================================

    **pandas** is a Python package providing fast, flexible,␣
↪and expressive data
    structures designed to make working with "relational" or
↪"labeled" data both
    easy and intuitive. It aims to be the fundamental high-
↪level building block for
    doing practical, **real world** data analysis in Python.␣
↪Additionally, it has
    the broader goal of becoming **the most powerful and␣
↪flexible open source data
    analysis / manipulation tool available in any language**.␣
↪It is already well on
    its way toward this goal.

    Main Features
    -------------
    Here are just a few of the things that pandas does well:

      - Easy handling of missing data in floating point as␣
↪well as non-floating
        point data.
      - Size mutability: columns can be inserted and deleted␣
↪from DataFrame and
```

(continues on next page)

(continued from previous page)

```
            higher dimensional objects
        - Automatic and explicit data alignment: objects can be␣
→explicitly aligned
        to a set of labels, or the user can simply ignore the␣
→labels and let
            `Series`, `DataFrame`, etc. automatically align the␣
→data for you in
        computations.
        - Powerful, flexible group by functionality to perform␣
→split-apply-combine
        operations on datasets, for both aggregating and␣
→transforming data.
        - Make it easy to convert ragged, differently-indexed␣
→data in other Python
        and NumPy data structures into DataFrame objects.
        - Intelligent label-based slicing, fancy indexing, and␣
→subsetting of large
        datasets.
        - Intuitive merging and joining datasets.
        - Flexible reshaping and pivoting of datasets.
        - Hierarchical labeling of axes (possible to have␣
→multiple labels per tick).
        - Robust IO tools for loading data from flat files (CSV␣
→and delimited),
        Excel files, databases, and saving/loading data from␣
→the ultrafast HDF5
        format.
        - Time series-specific functionality: date range␣
→generation and frequency
        conversion, moving window statistics, date shifting␣
→and lagging.

PACKAGE CONTENTS
    _config (package)
    _libs (package)
    _testing
    _typing
    _version
    api (package)
    arrays (package)
    compat (package)
    conftest
    core (package)
    errors (package)
    io (package)
    plotting (package)
    testing
    tests (package)
    tseries (package)
    util (package)
```

(continues on next page)

(continued from previous page)

```
SUBMODULES
    _hashtable
    _lib
    _tslib
    offsets

FUNCTIONS
    __getattr__ (name)

DATA
    IndexSlice = <pandas.core.indexing._IndexSlice object>
    NA = <NA>
    NaT = NaT
    __docformat__ = 'restructuredtext'
    __git_version__ = 'db08276bc116c438d3fdee492026f8223584c477
↪'
    describe_option = <pandas._config.config.
↪CallableDynamicDoc object>
    get_option = <pandas._config.config.CallableDynamicDoc␣
↪object>
    options = <pandas._config.config.DictWrapper object>
    reset_option = <pandas._config.config.CallableDynamicDoc␣
↪object>
    set_option = <pandas._config.config.CallableDynamicDoc␣
↪object>

VERSION
    1.1.3

FILE
    c:\users\zheng_\anaconda3_\envs\qe-mini-example\lib\site-
↪packages\pandas\__init__.py
```

Similar to docstring, as the list of functions and attributes in pandas can be overwhelmingly huge. We would truncate to the top 15 in the library. To help with development, we can always open another shell and open python in interactive mode and try out some of the functions and atrribute in different modules.

```
dir(pd)[:15]
```

```
['BooleanDtype',
 'Categorical',
 'CategoricalDtype',
 'CategoricalIndex',
 'DataFrame',
 'DateOffset',
 'DatetimeIndex',
 'DatetimeTZDtype',
 'ExcelFile',
 'ExcelWriter',
```

(continues on next page)

(continued from previous page)

```
'Float64Index',
'Grouper',
'HDFStore',
'Index',
'IndexSlice']
```

Also, notice that the directory is sorted in alphabetical order. To look for any variables, we can always do a list comprehension and filter. For example, to look for all the attributes starting with "d," we can:

```
[i for i in dir(pd) if i.lower().startswith('d')]
```

```
['DataFrame',
 'DateOffset',
 'DatetimeIndex',
 'DatetimeTZDtype',
 'date_range',
 'describe_option']
```

Alternatively, we can use filter to help us to get to the attributes we are looking for:

```
list(filter(lambda x: x.lower().startswith('d'), dir(pd)))
```

```
['DataFrame',
 'DateOffset',
 'DatetimeIndex',
 'DatetimeTZDtype',
 'date_range',
 'describe_option']
```

5.5 Writing Packages

Packages are libraries containing multiple modules and files. They are stored in directories and have one important requirement: each package is a directory which **must** contain an initialisation file called (unsurprisingly) __init__.py.

The file can be entirely empty, but it is imported and executed with the import function. This permits you to set some rules or initial steps to be performed with the first importation of the package.

You may be concerned that—with the modular nature of Python files and code— you may import a single library multiple times. Python keeps track and will only import (and initialize) the package once.

One useful part of the __init__.py file is that you can limit what is imported with the command from package import *.

```
#__init__.py

__all__ = ["class1", "class2"]
```

This means that `from package import *` actually only imports `class1` and `class2`

The next two sections are optional since, at this stage of your development practice, you are far less likely to need to produce code of this nature, but it can be useful to see how Python can be used in a slightly more advanced way.

5.6   Closures

Python has the concept of `scopes`. The variables created within a class or a function are only available within that class or function. The variables are available within the `scope` of the place they are called. If you want variables to be available within a function, you pass them as arguments (as you have seen previously).

Sometimes you want to have a global argument available to all functions, and sometimes you want a variable to be available to specific functions without being available more generally. Functions that can do this are called `closures`, and closures start with `nested functions`.

A `nested function` is a function defined inside another function. These nested functions gain access to the variables created in the enclosing scope.

```
def transmit_to_space(message):
    """
    This is the enclosing function
    """
    def data_transmitter():
        """
        The nested function
        """
        print(message)
    # Now the enclosing function calls the nested function
    data_transmitter()

transmit_to_space("Test message")
```

```
Test message
```

It is useful to remember that functions are also objects, so we can simply return the nested function as a response.

```
def transmit_to_space(message):
    """
    This is the enclosing function
    """
```

(continues on next page)

(continued from previous page)

```
    def data_transmitter():
        """
        The nested function
        """
        print(message)
    # Return an object of the nested function (i.e. without
↪brackets)
    return data_transmitter

msg = transmit_to_space("Into the sun!")
msg()
```

```
Into the sun!
```

5.7 Decorators

Closures may seem a little esoteric. Why would you use them?

Think in terms of the modularity of Python code. Sometimes you want to pre-process arguments before a function acts on them. You may have multiple different functions, but you want to validate your data in the same way each time. Instead of modifying each function, it would be better to enclose your function and only return data once your closure has completed its task.

One example of this is in websites. Some functions should only be executed if the user has the rights to do so. Testing for that in every function is tedious.

Python has syntax for enclosing a function in a closure. This is called the `decorator`, which has the following form:

```
@decorator
def functions(arg):
    return True
```

This is equivalent to `function = decorator(function)`, which is similar to the way the closures are structured in the previous section.

As a silly example, we have

```
def repeater(old_function):
    """
    A closure for any function which, passed as `old_function`
    returns `new_function`
    """
    def new_function(*args, **kwds):
        """
        A demo function which repeats any function in the
↪outer scope.
        """
```

(continues on next page)

(continued from previous page)

```
            old_function(*args, **kwds)
            old_function(*args, **kwds)
    return new_function

# We user `repeater` as a decorator like this
@repeater
def multiply(num1, num2):
    print(num1 * num2)

# And execute
multiply(6,7)
```

```
42
42
```

You can modify the output as well as the input.

```
def exponent_out(old_function):
    """
    This modification works on any combination of args and␣
↪kwargs.
    """
    def new_function(*args, **kwargs):
        return old_function(*args, **kwargs) ** 2
    return new_function

def exponent_in(old_function):
    """
    This modification only works if we know we have one␣
↪argument.
    """
    def new_function(arg):
        return old_function(arg ** 2)
    return new_function

@exponent_out
def multiply(num1, num2):
    return num1 * num2

print(multiply(6,7))

@exponent_in
def digit(num):
    return num

print(digit(6))

# And, let's trigger an error
@exponent_in
def multiply(num1, num2):
    return num1 * num2
```

(continues on next page)

(continued from previous page)

```
print(multiply(6,7))
```

```
1764
36
```

```
--------------------------------------------------------------
↪------------
TypeError                              Traceback (most␣
↪recent call last)
<ipython-input-15-2f68e5a397d0> in <module>()
     32         return num1 * num2
     33
---> 34 print(multiply(6,7))

TypeError: new_function() takes 1 positional argument but 2␣
↪were given
```

You can use decorators to check that an argument meets certain conditions before running the function.

```
class ZeroArgError(Exception):
    pass

def check_zero(old_function):
    """
    Check the argument passed to a function to ensure it is␣
↪not zero.
    """
    def new_function(arg):
        if arg == 0:
            raise ZeroArgError ("Zero is passed to argument")
        old_function(arg)
    return new_function

@check_zero
def print_num(num):
    print(num)

print_num(0)
```

```
--------------------------------------------------------------
↪------------
ZeroArgError                           Traceback (most␣
↪recent call last)
<ipython-input-22-b35d37f4e5e4> in <module>
     16         print(num)
     17
---> 18 print_num(0)
```

(continues on next page)

(continued from previous page)

```
<ipython-input-22-b35d37f4e5e4> in new_function(arg)
      8       def new_function(arg):
      9           if arg == 0:
---> 10               raise ZeroArgError ("Zero is passed to
↪argument")
     11           old_function(arg)
     12       return new_function

ZeroArgError: Zero is passed to argument
```

Sometimes, though, you want to pass new arguments to a decorator so that you can do something before executing your function. That rests on doubly nested functions.

```
def multiply(multiplier):
    """
    Using the multiplier argument, modify the old function to
↪return
    multiplier * old_function
    """
    def multiply_generator(old_function):
        def new_function(*args, **kwds):
            return multiplier * old_function(*args, **kwds)
        return new_function
    return multiply_generator

@multiply(3)
def return_num(num):
    return num

return_num(5)
```

```
15
```

Chapter 6
Advanced Python

Abstract Advanced Python will cover concepts in Python to allow a deeper understanding of its behavior. Topics covered are Magic Methods, Comprehension, Functional Parts, Iterables, Decorators, Object Oriented Programming, Properties, and Metaclasses. Magic methods are special methods that can enrich class designs by giving access to Python built-in syntax features. Comprehension, Functional Parts, Iterables, and Decorators are useful features that can help make code writing simpler and cleaner. Learning about Properties and Metaclasses work helps write better structured code using the Object Oriented Programming Paradigm.

Learning outcomes:

- Understand `python` magic methods and what is Pythonic code.
- Learn and apply object oriented concepts in `python`.
- Understand how MRO is done in `python`.
- Explore advanced tips and tricks.

Now we will try to cover some advanced features of the language.

6.1 Python Magic Methods

Before we look at magic methods, here is a quick overview of the different types of method naming conventions:

```
1. _method      : To prevent automatic import due to a "from xyz␣
↪import *" statement.
2. __method     : To mark as a private method.
3. method_      : To deal with reserved words
4. __method__   : magic functions, hooks that are triggered on␣
↪various builtin operators and functions.
```

© The Author(s), under exclusive license to Springer Nature Singapore Pte Ltd. 2022
T. T. Teoh, Z. Rong, *Artificial Intelligence with Python*,
Machine Learning: Foundations, Methodologies, and Applications,
https://doi.org/10.1007/978-981-16-8615-3_6

```
class A(object):
    def __foo(self):
        print("A foo")
    def class_(self):
        self.__foo()
        print(self.__foo.__name__)
    def __doo__(self):
        print("doo")

a = A() # instantiate A and assign to object a
print(hasattr(a, '__foo')) # where has this gone?
a.class_()
a.__doo__()
```

```
False
A foo
__foo
doo
```

So we can see here that we cannot access the private method outside the class, this is due to name mangling. The members can be inspected using the built-in `dir` method.

```
print(dir(a))
```

```
['_A__foo', '__class__', '__delattr__', '__dict__', '__dir__',
 ↪'__doc__', '__doo__', '__eq__', '__format__', '__ge__', '__
 ↪getattribute__', '__gt__', '__hash__', '__init__', '__init_
 ↪subclass__', '__le__', '__lt__', '__module__', '__ne__', '__
 ↪new__', '__reduce__', '__reduce_ex__', '__repr__', '__
 ↪setattr__', '__sizeof__', '__str__', '__subclasshook__', '__
 ↪weakref__', 'class_']
```

As we see, the name has changed to _A_foo and that is the only reason it is "private"; if we explicitly call it by its mangled name, it is very much accessible.

```
a._A__foo()
```

```
A foo
```

It is worth noting that the `__magic_method__` format does not do anything special unless we use the predefined names. It is also strongly advised that you only override the built-in magic methods and not redefine your own as I had just done previously.

```
class P(object):
    def __init__(self, x, y):
        self.x = x
        self.y = y
```

(continues on next page)

(continued from previous page)

```python
    def __add__(self, other):
        return P(self.x + other.x, self.y + other.y)

    def __gt__(self, other):
        #if both x and y component is greater than the other
↪object's x and y
        return (self.x > other.x) and (self.y > other.y)

    def __str__(self):
        return "x : %s, y : %s" % (self.x, self.y)

p1 = P(0,0)
p2 = P(3,4)
p3 = P(1,3)

print(p3 + p2)
print(p1 > p2)
print(p2 > p1)
```

```
x : 4, y : 7
False
True
```

You can even add stuff like **slicing capabilities**

```python
class Seq(object):
    def __getitem__(self, i):
        if type(i) is slice:
            # this has edge case issues, but just a demo!
            return list(range(i.start, i.stop))
        else:
            return i

s = Seq()
print(s[5])
print(s[-4])
print(s[2:5])
```

```
5
-4
[2, 3, 4]
```

Note We have covered a very small subset of all the "magic" functions. Please do have a look at the official Python docs for the exhaustive reference.

6.1.1 *Exercise*

In this exercise, we will practice how to use magic method in Python.

Task: The exchange rate between SGD and Euro is 1 to 1.8. Do some arithmetic operations with SGD and Euro.

- Add 20 SGD to 40 EURO and give your answer in SGD.
- Subtract 20 SGD from 40 EURO and give your answer in SGD.
- Subtract 20 Euro from 100 SGD and give your answer in Euro.

```
## First way

print(20 + 40 * 1.8)
print(40 * 1.8 - 20)
print(100/1.8 - 20)
```

```
92.0
52.0
35.55555555555556
```

The answer does not look good.

A clearer and more understandable approach is to use OOP.

Task: **Implement the following tasking using class and OO.**

6.1.2 *Solution*

```
exchange = {"SGD":{"Euro":1.8}}

class Money:
    def __init__(self, amount,currency):
        self.amt =amount
        self.currency = currency

    def __add__(self, money):
        if isinstance(money,self.__class__):
            if self.currency == money.currency:
                return Money(self.amount+money.amt,self.
↪currency)
            else:
                converted_rate = exchange[self.currency][money.
↪currency]*money.amt
                return Money(self.amt+converted_rate,self.
↪currency)

    def __sub__(self,money):
        money.amt*=-1
        return self.__add__(money)
```

(continues on next page)

(continued from previous page)

```
    def convert(self, currency):
        converted_rate = exchange[self.
↪currency][currency]*self.amt
        return Money(converted_rate,currency)

    def __repr__(self):
        return "The amount is {} in {}".format(round(self.amt,
↪2),self.currency)
```

```
Money(20,'SGD') + Money(40,'Euro')
```

```
The amount is 92.0 in SGD
```

```
Money(20,'SGD') - Money(10, 'Euro')
```

```
The amount is 2.0 in SGD
```

```
m = Money(100,'SGD') - Money(20, 'Euro')
m.convert('Euro')
```

```
The amount is 115.2 in Euro
```

6.2 Comprehension

Python comprehensions give us interesting ways to populate built-in data structures, *in terms of expressions*, as mathematicians would do. Comprehensions are a paradigm borrowed from functional languages and provide a great deal of syntactic sugar.

```
l = [i for i in range(0,5)]
l2 = [i*i for i in range(0,5)]
```

We can also define slightly more complex expressions with the use of if statements and nested loops.

```
l = [i for i in range(0,5) if i % 2 ==0]
print(l)
```

```
[0, 2, 4]
```

```
# get all combinations where x > y and x, y < 5
xy = [ (x, y) for x in range (0,5) for y in range (0, 5) if x >
↪ y]
print(xy)
```

```
[(1, 0), (2, 0), (2, 1), (3, 0), (3, 1), (3, 2), (4, 0), (4,␣
↪1), (4, 2), (4, 3)]
```

```
# we can even call functions
l = [x.upper() for x in "hello"]
print(l)
```

```
['H', 'E', 'L', 'L', 'O']
```

```
# creating lists of lists is also a synch
gre = "hello how are you doing?"
[[s.lower(), s.upper(), len(s)] for s in gre.split()]
```

```
[['hello', 'HELLO', 5],
 ['how', 'HOW', 3],
 ['are', 'ARE', 3],
 ['you', 'YOU', 3],
 ['doing?', 'DOING?', 6]]
```

```
# nested comprehensions - we can do it, but it may not be very␣
↪readable
matrix = [[i+x for i in range(3)] for x in range(3)]
print (matrix)
```

```
[[0, 1, 2], [1, 2, 3], [2, 3, 4]]
```

```
# we can also have a comprehension for dicts
d = {x : x**2 for x in range(5)}
print (d)
```

```
{0: 0, 1: 1, 2: 4, 3: 9, 4: 16}
```

6.3   Functional Parts

There are a lot of concepts borrowed from functional languages to make the code look more elegant. List comprehension was just scratching the surface. We have built-in helpers such as lambda, filter, zip, map, all, any to help us write cleaner code. Other than the built-in components, we have functools (which I will not be covering) which even helps us with partial functions and currying.

```
# lambda is used when you need anonymous functions defined as
↪an expression
# in this example you could define a function and pass it to
↪foo, or use the lambda
# in this case the lambda is neater.
# lambdas can take in n number of params, and the body is a
↪single expression that is also the return value

def foo(list_, func):
    l = []
    for i in list_:
        l.append(func(i))
    return l

def sq(i):
    return i**2

l = [i for i in range(5)]
print (foo(l, sq))
print (foo(l, lambda x : x**2))
```

```
[0, 1, 4, 9, 16]
[0, 1, 4, 9, 16]
```

```
class P(object):
    def __init__(self, x,y):
        self.x = x
        self.y = y
    def __str__(self):
        return "x : %s" % self.x

l = [P(5,5), P(2,2), P(1,1), P(4,4), P(3,3)]
l.sort(key=lambda x: (x.x)**2 + (x.y)**2)
for p in l : print (p) # [str(p) for p in l]

# there are many more complex and cryptic ways to use
↪ (exploit) lambdas,
#     you can search for it online if you are interested
# check lambda with multiple args
# lambda *x : sys.stdout.write(" ".join(map(str, x)))
```

```
x : 1
x : 2
x : 3
x : 4
x : 5
```

```
# filter is a function that takes an interable and a callable
#  applies the function to each element, i.e., ret =
↪func(element)
```

(continues on next page)

(continued from previous page)

```
#  and returns a list with elements for which 'ret' was true

l = range(0,10)
l = filter(lambda x : x%2==0, l)
print (l, type(l))
```

```
<filter object at 0x000001EDD08B9548> <class 'filter'>
```

```
# zip is to sew together a bunch of iterables
# the list generated is of the minimum size of all the
↪iterators that have gone in!

a = [1,2,3,4,5]
b = (0,4,6,7)
c = {1:'a', 7:'b', 'm':'v'}

print (zip(a,b,c))
```

```
<zip object at 0x000001EDD0755C48>
```

Though zip looks trivial, it is a fairly important operation for mathematical algorithms—matrices, curve fitting, interpolation, pattern recognition, that sort of thing. It is also very important in engineering applications like digital signal processing where much of what you do is combine multiple signals or apply linear transforms to them—both are based on the sample index, hence, zip it.

It would be a pain to reimplement it every time taking care of all edge cases, etc.

```
# map - takes in an iterable and callable - applies the
↪callable to each element of the iterable
#  returns a new list with each element being the return value
↪of "callable(elem)"

print (map(lambda x: x**2, range(10)))
```

```
<map object at 0x000001EDD0755108>
```

```
# map is extremely useful as a shorthand for "applying" a
↪function across an iterable,
#  especially in conjunction with lambda
import sys
my_print = lambda *x : sys.stdout.write(" ".join(map(str, x)))
my_print("hello", "how are you", 1234)
```

```
hello how are you 1234
```

6.4   Iterables

Lists, dicts, and tuples are iterables. That is, we can "iterate" through them. Any object that supports the iterator protocol is an iterator. The iterator protocol states that the object should override the __iter__ magic method that returns an object that has a .next() method and raises a StopIteration exception.

There are 4 key ways to create an iterable:

```
1. Iterators - classes that override __iter__ and next()
2. Generator functions - functions that yield
3. Generator expressions
4. overriding the __getitem__ magic method.
```

```
# 1. Iterators

class myitr:
    def __init__(self, upper_limit=5):
        self.limit=upper_limit
    def __iter__(self):
        self.index = 0
        return self
    def __next__(self):
        if self.index < self.limit:
            self.index += 1
            return self.index
        else:
            raise StopIteration

for i in myitr(5):
    print(i)
```

```
1
2
3
4
5
```

```
# 2. Generators

def gen(lim):
    i = 0
    while i < lim:
        yield i
        i = i + 1

for i in gen(5):
    print(i)
```

```
0
1
2
3
4
```

```
# 3. Generator expression

def seq(num):
    return (i**2 for i in range(num))

for i in seq(5):
    print(i)
```

```
0
1
4
9
16
```

```
# 4. Overriding __getitem__

class Itr(object):
    def __init__(self, x):
        self.x = x

    def __getitem__(self, index):
        if index < self.x:
            return index
        else:
            raise StopIteration


for i in Itr(5):
    print(i)
```

```
0
1
2
3
4
```

6.5 Decorators

Before we start with decorators, we need to know a bit about closures. A closure is a function object that remembers values in enclosing scopes regardless of whether those scopes are still present in memory. The most common case is when we define a function within a function and return the inner function. If the inner function definition uses variables/values in the outer function, it maintains the references to those even after it is returned (and no longer in the scope of the outer function).

```
# closure example - raised_to_power returns a fn that takes a␣
↪variable and raises to the power 'n'
#  'n' is passed only once - while defining the function!

def raised_to_power(n):
    def fn(x):
        return x**n
    return fn

p2 = raised_to_power(2)
p3 = raised_to_power(3)

print (p2(2), p2(3)) # still remembers that n=2
print (p3(2), p3(3)) # still remembers that n=3
```

```
4 9
8 27
```

```
# have to be cautious!

def power_list(n):
    '''returns list of fn, each raises to power i, where i : 0␣
↪--> n'''
    fn_list = []

    def fn(x):
        return x**i

    for i in range(n):
        # doesn't matter if fn was defined here either
        fn_list.append(fn)

    return fn_list

for j in power_list(4):
    print (j(2)) # prints 2 power 3, 4 times
```

```
8
8
8
8
```

```
# decorator is just a nicer way of defining a closure - more
↪syntactic sugar

def deco(fn):
    def new_fn(*args, **kwargs):
        print ("entering function", fn.__name__)
        ret = fn(*args, **kwargs)
        print ("exiting function", fn.__name__)
    return new_fn

@deco
def foo(x):
    print("x : ", x)

foo(4)
```

```
entering function foo
x :   4
exiting function foo
```

```
# Another example

def add_h1(fn):
    def nf(pram):
        return "<h1> " + fn(pram) + " </h1>"
    return nf

@add_h1
def greet(name):
    return "Hello {0}!".format(name)

print greet("Nutanix")
```

```
<h1> Hello Nutanix! </h1>
```

```
# decorator that takes parameter

def add_h(num):
    def deco(fn):
        # this is the decorator for a specific 'h'
        def nf(pram):
            return "<h%s> "%num + fn(pram) + " </h%s>"%num
        return nf
    return deco

@add_h(3)
def greet(name):
    return "Hello {0}!".format(name)
print (greet("Nutanix"))
```

(continues on next page)

(continued from previous page)

```python
# we can have multiple decorators as well
@add_h(2)
@add_h(4)
def greet2(name):
    return "Hello {0}!".format(name)

print (greet2("Nutanix"))
```

```
<h3> Hello Nutanix! </h3>
<h2> <h4> Hello Nutanix! </h4> </h2>
```

6.6 More on Object Oriented Programming

Let us take another look at classes and OO in python. We will start with multiple inheritance (or mixins).

6.6.1 Mixins

Let us start with this inheritance model:

```
    A
   / \
  B   C
   \ /
    D
```

```python
class A(object):
    def __init__(self):
        print ("A.init")
    def foo(self):
        print ("A.foo")

class B(A):
    def __init__(self):
        print ("B.init")
    def foo(self):
        print ("B.foo")

class C(A):
    def __init__(self):
        print ("C.init")
```

(continues on next page)

(continued from previous page)

```
    def foo(self):
        print ("C.foo")

class D(B, C):
    def __init__(self):
        print ("D.init")
    #def foo(self):
    #    print "D.foo"

class E(C, B):
    def __init__(self):
        print ("E.init")

d = D()
d.foo()

e = E()
e.foo()

# we see that fn lookup's happen in the order of declaration␣
↪of parent in the child's definition.
```

```
D.init
B.foo
E.init
C.foo
```

What if the mixin is slightly more complex? (Note, no matter how complex stuff gets—which it should not, Python will never let you create a circular dependency!)

```
      A
    /   \
   B    C
   |  / |
   D/   |
   |    |
    \   |
      E
```

```
class A(object):
    def __init__(self):
        print ("A.init")
    def foo(self):
        print ("A.foo")

class B(A):
    def __init__(self):
        print ("B.init")
    def foo(self):
        print ("B.foo")
```

(continues on next page)

(continued from previous page)

```
class C(A):
    def __init__(self):
        print ("C.init")
    def foo(self):
        print ("C.foo")

class D(C):
    def __init__(self):
        print ("D.init")
    def foo(self):
        print ("D.foo")

class E(D, C): # you can't have (C, D) - TypeError: Cannot␣
↪create a consistent MRO
    def __init__(self):
        print ("E.init")

e = E()
e.foo()
E.__mro__

# so what's mro - (explain in live session)
```

```
E.init
D.foo
```

```
(__main__.E, __main__.D, __main__.C, __main__.A, object)
```

Note MRO is also the reason why super() is called in the manner it is. You need both the class and the object to traverse the next parent in the MRO.

6.6.2 Attribute Access Hooks

Next let us have a look at two magic functions which deal with object variable access, __getattr__ and __setattr__. The __getattr__ method returns the value of the named attribute of an object. If not found, it returns the default value provided to the function. __setattr__ is called when an attribute assignment is attempted.

This allows us to hook into attribute setting and assignment conveniently.

```
class A(object):
    def __init__(self, x):
        self.x = x
    def __getattr__(self, val):
```

(continues on next page)

(continued from previous page)

```
            print ("getattr val :", val, type(val))
            return val

a = A(3)
print "X :", a.x # getattr not called for x
ret = a.y
print ("Y :", ret)
```

```
X : 3
getattr val : y <type 'str'>
Y : y
```

Here are some uses cases. __getattr__ can help us refactor and clean up our code. This is handy in lots of places and avoids having to wrap things in try/except blocks. Consider the following:

```
class settings:
    pass
try:
    foo = settings.FOO
except AttributeError:
    foo = None
```

The code can be replaced by
foo = getattr(settings, 'FOO', None)

```
class A(object):
    def __init__(self, x):
        self.x = x
    def __getattr__(self, val):
        print ("getattr")
        return val
    def __setattr__(self, name, val):
        print ("setattr")
        if name == 'x':
            self.__dict__[name] = val

a = A(3)
print (a.x)
print (a.y)

# setattr is called for both
a.y = 5
a.x = 5
```

```
setattr
3
getattr
y
setattr
setattr
```

6.6.3   Callable Objects

You can make an object callable (a functor) by overriding the magic __call__ method. You can call the object like a function and the __call__ method will be called instead. This is useful when you want to have more complex functionality (like state) plus data but want to keep the syntactic sugar/simplicity of a function.

```python
class MulBy(object):
    def __init__(self, x):
        self.x = x
    def __call__(self, n):
        print ("here!")
        return self.x * n

m = MulBy(5)
print (m(3))
```

```
here!
15
```

6.6.4   _new_ vs _init_

Until now we never bothered to see how/when the Python objects were created. The __init__ function just deals with handling the initialization of the object, and the actual creation happens within __new__, which can be overridden.

From the Python mailing list, we have

```
Use __new__ when you need to control the creation of a new
↪instance.
Use __init__ when you need to control initialization of a new
↪instance.

__new__ is the first step of instance creation.  It's called
↪first,
and is responsible for returning a new instance of your class.
↪ In
```

(continues on next page)

(continued from previous page)

```
contrast, __init__  doesn't return anything; it's only␣
↪responsible for
initializing the instance after it's been created.
```

```
class X(object):
    def __new__(cls, *args, **kwargs):
        print ("new")
        print (args, kwargs)
        return object.__new__(cls)

    def __init__(self, *args, **kwargs):
        print ("init")
        print (args, kwargs)

x = X(1,2,3,a=4)
```

```
new
(1, 2, 3) {'a': 4}
init
(1, 2, 3) {'a': 4}
```

This approach is useful for the factory design pattern.

```
class WindowsVM(object):
    def __init__(self, state="off"):
        print ("New windows vm. state : %s" %state)
    def operation(self):
        print ("windows ops")

class LinuxVM(object):
    def __init__(self, state="off"):
        print ("New linux vm. state : %s" %state)
    def operation(self):
        print ("linux ops")

class VM(object):
    MAP = {"Linux" : LinuxVM, "Windows": WindowsVM}

    def __new__(self, vm_type, state="off"):
        # return object.__new__(VM.MAP[vm_type]) #--doesn't␣
↪call init of other class
        vm = object.__new__(VM.MAP[vm_type])
        vm.__init__(state)
        return vm


vm1 = VM("Linux")
print (type(vm1))
vm1.operation()
print ()
```

(continues on next page)

(continued from previous page)

```
vm2 = VM("Windows", state="on")
print (type(vm2))
vm2.operation()
```

```
New linux vm. state : off
<class '__main__.LinuxVM'>
linux ops

New windows vm. state : on
<class '__main__.WindowsVM'>
windows ops
```

6.7 Properties

Properties are ways of adding behavior to instance variable access, i.e., trigger a function when a variable is being accessed. This is most commonly used for getters and setters.

```
# simple example
class C(object):
    def __init__(self):
        self._x = None

    def getx(self):
        print ("getx")
        return self._x

    def setx(self, value):
        print ("setx")
        self._x = value

    def delx(self):
        print ("delx")
        del self._x

    x = property(getx, setx, delx, "I'm the 'x' property.")

c = C()
c.x = 5 # so when we use 'x' variable of a C object, the
↪getters and setters are being called!
print (c.x)
del c.x

#print (C.x) error
```

```
setx
getx
5
delx
```

```
# the same properties can be used in form of decorators!
class M(object):
    def __init__(self):
        self._x = None

    @property
    def x(self):
        print ("getx")
        return self._x

    @x.setter
    def x(self, value):
        print ("setx")
        self._x = value

    @x.deleter
    def x(self):
        print ("delx")
        del self._x
m = C()
m.x = 5
print (m.x)
del m.x
```

```
setx
getx
5
delx
```

So how does this magic happen? how do properties work? It so happens that
properties are data descriptors. Descriptors are objects that have a __get__,
__set__, __del__ method. When accessed as a member variable, the corresponding function gets called. Property is a class that implements this descriptor
interface, there is nothing more to it.

```
# This is a pure python implementation of property

class Property(object):
    "Emulate PyProperty_Type() in Objects/descrobject.c"

    def __init__(self, fget=None, fset=None, fdel=None, ↵
↪doc=None):
        self.fget = fget
        self.fset = fset
```

(continues on next page)

(continued from previous page)

```
            self.fdel = fdel
            if doc is None and fget is not None:
                doc = fget.__doc__
            self.__doc__ = doc

    def __get__(self, obj, objtype=None):
        if obj is None:
            return self
        if self.fget is None:
            raise AttributeError("unreadable attribute")
        return self.fget(obj)

    def __set__(self, obj, value):
        if self.fset is None:
            raise AttributeError("can't set attribute")
        self.fset(obj, value)

    def __delete__(self, obj):
        if self.fdel is None:
            raise AttributeError("can't delete attribute")
        self.fdel(obj)

    def getter(self, fget):
        return type(self)(fget, self.fset, self.fdel, self.__
↪doc__)

    def setter(self, fset):
        return type(self)(self.fget, fset, self.fdel, self.__
↪doc__)

    def deleter(self, fdel):
        return type(self)(self.fget, self.fset, fdel, self.__
↪doc__)
# during the live session, explain how this maps to the␣
↪previous decorator version of property.
```

6.8 Metaclasses

Metaclasses are classes that create new classes (or rather a class whose instances are classes themselves). They are useful when you want to dynamically create your own types. For example, when you have to create classes based on a description file (XML)—like in the case of some libraries built over WSDL (PyVmomi) or in the case when you want to dynamically mix two or more types of classes to create a new one (e.g., a VM type, an OS type, and an interface type—used in NuTest framework developed by the automation team).

Another reason is to enforce some checking/have restrictions in the user-defined classes. Like in the case of Abstract Base Class (ABCMeta). The metaclass that creates the user-defined class can run some pre-checks (whether certain functions are defined, etc.) and some preprocessing (adding new methods, etc.) if required.

```
class MyMet(type):
    """Here we see that MyMet doesn't inherit 'object' but
↪rather 'type' class - the builtin metaclass
    """
    def __new__(cls, name, bases, attrs):
        """
        Args:
            name (str) : name of the new class being created
            bases (tuple) : tuple of the classes which are the
↪parents of cls
            attrs (dict) : the attributes that belong to the
↪class
        """
        print ("In new")
        print (name)
        print (bases)
        print (attrs)
        return super(MyMet, cls).__new__(cls, name, bases,
↪attrs)

    def __init__(self, *args, **kwargs):
        print ("In init")
        print (self)
        print (args)
        print (kwargs)

class Me(object):
    __metaclass__ = MyMet

    def foo(self):
        print ("I'm foo")


m = Me()
m.foo()
```

```
I'm foo
```

In this case, we see that "m" which is an instance of "Me" works as expected. Here we are using the metaclass to just print out the flow, but we can do much more.

Also if you note, we see that the args to __init__ are the same as args to __new__, which is again as expected.

```
class MyMet(type):
    """Here we see that MyMet doesn't inherit 'object' but
↪rather 'type' class - the builtin metaclass
    """
```

(continues on next page)

(continued from previous page)

```python
    def __new__(cls, name, bases, attrs):
        """
        Args:
          name (str) : name of the new class being created
          bases (tuple) : tuple of the classes which are the
→parents of cls
          attrs (dict) : the attributes that belong to the
→class
        """
        print ("In new")
        print (name)
        print (bases)
        print (attrs)
        def foo(self):
            print ("I'm foo")
        attrs['foo'] = foo
        return super(MyMet, cls).__new__(cls, name, bases,
→attrs)

    def __init__(self, name, bases, attrs):
        print ("In init")
        print (self) # actually the object being created
        print (name)
        print (bases)
        print (attrs)
        def bar(self):
            print ("I'm bar")
        setattr(self, "bar", bar)

    def test(self):
        print ("in test")

    #def __call__(self):
    #    print "self :", self
    # Note : If I override call here, then I have to
→explicitly call self.__new__
    #        otherwise it is completely skipped. Normally a
→class calls type's __call__
    #        which re-routes it to __new__ of the class


class Me(object):
    __metaclass__ = MyMet

    def foo(self):
        pass

    def bar(self):
        pass
```

(continues on next page)

(continued from previous page)

```
print ("\n---------------------------\n")

m = Me()
print (type(Me)) # not of type 'type' anymore!
m.foo()
m.bar()
print (type(m))
```

```
---------------------------

<class 'type'>
<class '__main__.Me'>
```

Note What the __metaclass__ does is it tells the interpreter to parse the class in question, get the name, the attribute dictionary, and the base classes, and create it using a "type" type, in this case, the MyMet class. In its most primitive form that is how classes are created, using the "type" inbuilt class. We use this a lot to dynamically mix classes in NuTest.

```
class A(object):
    def __init__(self):
        print ("init A")
    def foo(self):
        print ("foo A")
    def bar(self):
        print ("bar A")

class B(object):
    def __init__(self):
        print ("init B")
    def doo(self):
        print ("doo B")
    def bar(self):
        print ("bar B")

def test(self):
    print ("Self : ", self)

Cls = type("C", (A,B), {"test": test})

c = Cls()
print (Cls)
print (Cls.__name__, type(Cls))
print (c)
```

```
init A
<class '__main__.C'>
C <class 'type'>
<__main__.C object at 0x000002C3C0855448>
```

```
c.foo()
c.bar()
c.doo()
c.test()
```

```
foo A
bar A
doo B
Self :   <__main__.C object at 0x000002C3C0855448>
```

Chapter 7
Python for Data Analysis

Abstract This chapter will introduce Ethics in algorithm development and common tools that are used for data analysis. Ethics is an important consideration in developing Artificial Intelligence algorithms. The outputs of computers are derived from the data that are provided as input and the algorithms developed by Artificial Intelligence developers who must be held accountable in their analysis. Tools such as Numpy, Pandas, and Matplotlib are covered to aid in the data analysis process. Numpy is a powerful set of tools to work with complete data lists efficiently. Pandas is a package designed to make working with "relational" or "labeled" data both easy and intuitive. Finally, Matplotlib is a powerful Python plotting library used for data visualization.

Learning outcomes:

- Identify concepts in ethical reasoning, which may influence our analysis and results from data.
- Learn and apply a basic set of methods from the core data analysis libraries of Numpy, Pandas, and Matplotlib.

Data has become the most important language of our era, informing everything from intelligence in automated machines to predictive analytics in medical diagnostics. The plunging cost and easy accessibility of the raw requirements for such systems—data, software, distributed computing, and sensors—are driving the adoption and growth of data-driven decision-making.

As it becomes ever-easier to collect data about individuals and systems, a diverse range of professionals—who have never been trained for such requirements—grapple with inadequate analytic and data management skills, as well as the ethical risks arising from the possession and consequences of such data and tools.

Before we go on with the technical training, consider the following on the ethics of the data we use.

T. T. Teoh, Z. Rong, *Artificial Intelligence with Python*,
Machine Learning: Foundations, Methodologies, and Applications,
https://doi.org/10.1007/978-981-16-8615-3_7

7.1 Ethics

Computers cannot make decisions. Their output is an absolute function of the data provided as input, and the algorithms applied to analyze that input. The aid of computers in decision-making does not override human responsibility and accountability.

It should be expected that both data and algorithms should stand up to scrutiny so as to justify any and all decisions made as a result of their output. "Computer says no," is not an unquestionable statement.

Our actions—as data scientists—are intended to persuade people to act or think other than the way they currently do based on nothing more than the strength of our analysis, and informed by data.

The process by which we examine and explain why what we consider to be right or wrong is considered right or wrong in matters of human conduct belongs to the study of ethics.

Case-study: *Polygal* was a gel made from beet and apple pectin. Administered to a severely wounded patient, it was supposed to reduce bleeding. To test this hypothesis, Sigmund Rascher administered a tablet to human subjects who were then shot or—without anesthesia—had their limbs amputated.

During the Second World War, and under the direction of senior Nazi officers, medical experiments of quite unusual violence were conducted on prisoners of war and civilians regarded by the Nazi regime as sub-human. After the war, twenty medical doctors were tried for war crimes and crimes against humanity at the Doctor's Trial held in Nuremberg from 1946 to 1949.

In 1947 Kurt Blome—Deputy Reich Health Leader, a high-ranking Nazi scientist—was acquitted of war crimes on the strength of intervention by the USA. Within two months, he was being debriefed by the US military who wished to learn everything he knew about biological warfare.

Do you feel the USA was "right" or "wrong" to offer Blome immunity from prosecution in exchange for what he knew?

There were numerous experiments conducted by the Nazis that raise ethical dilemmas, including: immersing prisoners in freezing water to observe the result and test hypothermia revival techniques; high altitude pressure and decompression experiments; sulfanilamide tests for treating gangrene and other bacterial infections.

Do you feel it would be "right" or "wrong" to use these data in your research and analysis?

Whereas everything else we do can describe human behavior as it *is*, ethics provides a theoretical framework to describe the world as it *should*, or *should not* be. It gives us full range to describe an ideal outcome, and to consider all that we know and do not know, which may impede or confound our desired result.

Case-study: A Nigerian man travels to a conference in the USA. After one of the sessions, he goes to the bathroom to wash his hands. The electronic automated soap dispenser does not recognize his hands beneath the sensor. A white American sees his confusion and places his hands beneath the device. Soap is dispensed. The Nigerian man tries again. It still does not recognize him.

How would something like this happen? Is it an ethical concern?

When we consider ethical outcomes, we use the terms *good* or *bad* to describe judgments about people or things, and we use *right* or *wrong* to refer to the outcome of specific actions. Understand, though, that—while right and wrong may sometimes be obvious—we are often stuck in ethical dilemmas.

How we consider whether an action is right or wrong comes down to the tension between what was intended by an action, and what the consequences of that action were. Are only intensions important? Or should we only consider outcomes? And how absolutely do you want to judge this chain: the *right* motivation, leading to the *right* intention, performing the *right* action, resulting in only *good* consequences. How do we evaluate this against what it may be impossible to know at the time, even if that information will become available after a decision is made?

We also need to consider competing interests in good and bad outcomes. A good outcome for the individual making the decision may be a bad decision for numerous others. Conversely, an altruistic person may act only for the benefit of others even to their own detriment.

Ethical problems do not always require a call to facts to justify a particular decision, but they do have a number of characteristics:

- *Public*: the process by which we arrive at an ethical choice is known to all participants.
- *Informal*: the process cannot always be codified into law like a legal system.
- *Rational*: despite the informality, the logic used must be accessible and defensible.
- *Impartial*: any decision must not favor any group or person.

Rather than imposing a specific set of rules to be obeyed, ethics provides a framework in which we may consider whether what we are setting out to achieve conforms to our values, and whether the process by which we arrive at our decision can be validated and inspected by others.

No matter how sophisticated our automated machines become, unless our intention is to construct a society "of machines, for machines", people will always be needed to decide on what ethical considerations must be taken into account.

There are limits to what analysis can achieve, and it is up to the individuals producing that analysis to ensure that any assumptions, doubts, and requirements are documented along with their results. Critically, it is also each individual's personal responsibility to raise any concerns with the source data used in the analysis, including whether personal data are being used legitimately, or whether the source data are at all trustworthy, as well as the algorithms used to process those data and produce a result.

7.2   Data Analysis

This will be a very brief introduction to some tools used in data analysis in Python. This will not provide insight into the approaches to performing analysis, which is left to self-study, or to modules elsewhere in this series.

7.2.1   Numpy Arrays

Data analysis often involves performing operations on large lists of data. Numpy is a powerful suite of tools permitting you to work quickly and easily with complete data lists. We refer to these lists as arrays, and—if you are familiar with the term from mathematics—you can think of these as matrix methods.

By convention, we import Numpy as np: `import numpy as np`.

We're also going to want to be generating a lot of lists of random floats for these exercises, and that's tedious to write. Let's get Python to do this for us using the `random` module.

```python
import numpy as np
import random

def generate_float_list(lwr, upr, num):
    """
    Return a list of num random decimal floats ranged between
↪lwr and upr.

    Range(lwr, upr) creates a list of every integer between
↪lwr and upr.
    random.sample takes num integers from the range list,
↪chosen randomly.
    """
    int_list = random.sample(range(lwr, upr), num)
    return [x/100 for x in int_list]

# Create two lists
height = generate_float_list(100, 220, 10)
weight = generate_float_list(5000, 20000, 10)

# Convert these to Numpy arrays
np_height = np.array(height)
np_weight = np.array(weight)

print(np_height)
print(np_weight)
```

```
[1.54 1.75 1.43 2.03 1.51 1.59 1.19 1.72 1.13 2.09]
[ 70.08 166.31 170.51 174.34  89.29  69.13 137.76  96.66 123.
↪97  95.73]
```

There is a useful timer function built in to Jupyter Notebook. Start any line of code with %time and you'll get output on how long the code took to run.

This is important when working with data-intensive operations where you want to squeeze out every drop of efficiency by optimizing your code.

We can now perform operations directly on all the values in these Numpy arrays. Here are two simple methods to use.

- Element-wise calculations: you can treat Numpy arrays as you would individual floats or integers. Note, they must either have the same shape (i.e. number or elements), or you can perform bitwise operations (operate on each item in the array) with a single float or int.
- Filtering: You can quickly filter Numpy arrays by performing boolean operations, e.g. np_array[np_array > num], or, for a purely boolean response, np_array > num.

```
# Calculate body-mass index based on the heights and weights␣
↪in our arrays
# Time the calculation ... it won't be long
%time bmi = np_weight / np_height ** 2

print(bmi)

# Any BMI > 35 is considered severely obese. Let's see who in␣
↪our sample is at risk.

# Get a boolean response
print(bmi > 35)

# Or print only BMI values above 35
print(bmi[bmi > 35])
```

```
Wall time: 0 ns
[29.54967111 54.30530612 83.38305052 42.30629231 39.16056313␣
↪27.34464618
 97.28126545 32.67306652 97.08669434 21.91570706]
[False  True  True  True  True False  True False  True False]
[54.30530612 83.38305052 42.30629231 39.16056313 97.28126545␣
↪97.08669434]
```

7.2.2 Pandas

We briefly experimented with Pandas back in [Built-in modules](03 - Python intermediate.ipynb#Built-in-modules).

The description given for Pandas there was:

pandas *is a Python package providing fast, flexible, and expressive data structures designed to make working with "relational" or "labeled" data both easy and*

intuitive. It aims to be the fundamental high-level building block for doing practical,
***real world** data analysis in Python. Additionally, it has the broader goal of becoming*
the most powerful and flexible open source data analysis / manipulation tool
available in any language.

Pandas was developed by Wes McKinney and has a marvelous and active development community. Wes prefers Pandas to be written in the lower-case (I'll alternate).

Underneath Pandas is Numpy, so they are closely related and tightly integrated. Pandas allows you to manipulate data either as a `Series` (similarly to Numpy, but with added features) or in a tabular form with rows of values and named columns (similar to the way you may think of an Excel spreadsheet).

This tabular form is known as a `DataFrame`. Pandas works well with Jupyter Notebook and you can output nicely formatted dataframes (just make sure the last line of your code block is the name of the dataframe).

The convention is to import pandas as pd: `import pandas as pd`.

The following tutorial is taken directly from the "10 minutes to pandas" section of the Pandas documentation. Note, this isn't the complete tutorial, and you can continue there.

Object Creation in Pandas
The following code will be on object creation in Pandas.

Create a `Series` by passing a list of values, and letting pandas create a default integer index.

```
import pandas as pd
import numpy as np

s = pd.Series([1,3,5,np.nan,6,8])
s
```

```
0    1.0
1    3.0
2    5.0
3    NaN
4    6.0
5    8.0
dtype: float64
```

Note that `np.nan` is Numpy's default way of presenting a value as "not-a-number." For instance, divide-by-zero returns `np.nan`. This means you can perform complex operations relatively safely and sort out the damage afterward.

Create a DataFrame by passing a Numpy array, with a datetime index and labeled columns.

```
# Create a date range starting at an ISO-formatted date
↪ (YYYYMMDD)
dates = pd.date_range('20130101', periods=6)
dates
```

```
DatetimeIndex(['2013-01-01', '2013-01-02', '2013-01-03', '2013-
↪01-04',
               '2013-01-05', '2013-01-06'],
             dtype='datetime64[ns]', freq='D')
```

```
# Create a dataframe using the date range we created above as
↪the index
df = pd.DataFrame(np.random.randn(6,4), index=dates,
↪columns=list('ABCD'))
df
```

```
                    A          B          C          D
2013-01-01   1.175032  -2.245533   1.196393  -1.896230
2013-01-02   0.211655  -0.931049   0.339325  -0.991995
2013-01-03   1.541121   0.709584   1.321304   0.715576
2013-01-04  -0.180625  -1.332144  -0.503592  -0.458643
2013-01-05   1.024923  -1.356436  -2.661236   0.765617
2013-01-06  -0.209474  -0.739143   0.076423   2.346696
```

We can also mix text and numeric data with an automatically generated index.

```
dict = {"country": ["Brazil", "Russia", "India", "China",
↪"South Africa"],
        "capital": ["Brasilia", "Moscow", "New Delhi", "Beijing
↪", "Pretoria"],
        "area": [8.516, 17.10, 3.286, 9.597, 1.221],
        "population": [200.4, 143.5, 1252, 1357, 52.98] }

brics = pd.DataFrame(dict)
brics
```

```
          country     capital     area   population
0          Brazil    Brasilia    8.516       200.40
1          Russia      Moscow   17.100       143.50
2           India   New Delhi    3.286      1252.00
3           China     Beijing    9.597      1357.00
4    South Africa    Pretoria    1.221        52.98
```

The numbers down the left-hand side of the table are called the index. This permits you to reference a specific row. However, Pandas permits you to set your own index, as we did where we set a date range index. You could set one of the existing columns as an index (as long as it consists of unique values) or you could set a new custom index.

```
# Set the ISO two-letter country codes as the index
brics.index = ["BR", "RU", "IN", "CH", "SA"]

brics
```

```
              country     capital     area   population
BR             Brazil    Brasilia    8.516       200.40
RU             Russia      Moscow   17.100       143.50
IN              India   New Delhi    3.286      1252.00
CH              China     Beijing    9.597      1357.00
SA       South Africa    Pretoria    1.221        52.98
```

Pandas can work with exceptionally large datasets, including millions of rows. Presenting that takes up space and, if you only want to see what your data looks like (since, most of the time, you can work with it symbolically), then that can be painful. Fortunately, Pandas comes with a number of ways of viewing and reviewing your data.

- See the top and bottom rows of your dataframe with df.head() or df.tail(num) where num is an integer number of rows.
- See the index, columns, and underlying Numpy data with df.index, df.columns, and df.values, respectively.
- Get a quick statistical summary of your data with df.describe().
- Transpose your data with df.T.
- Sort by an axis with df.sort_index(axis=1, ascending=False) where axis=1 refers to columns, and axis=0 refers to rows.
- Sort by values with df.sort_values(by=column).

```
# Head
df.head()
```

```
                   A          B          C          D
2013-01-01  1.175032  -2.245533   1.196393  -1.896230
2013-01-02  0.211655  -0.931049   0.339325  -0.991995
2013-01-03  1.541121   0.709584   1.321304   0.715576
2013-01-04 -0.180625  -1.332144  -0.503592  -0.458643
2013-01-05  1.024923  -1.356436  -2.661236   0.765617
```

```
# Tail
df.tail(3)
```

```
                   A          B          C          D
2013-01-04 -0.180625  -1.332144  -0.503592  -0.458643
2013-01-05  1.024923  -1.356436  -2.661236   0.765617
2013-01-06 -0.209474  -0.739143   0.076423   2.346696
```

```
# Index
df.index
```

```
DatetimeIndex(['2013-01-01', '2013-01-02', '2013-01-03', '2013-
↪01-04',
               '2013-01-05', '2013-01-06'],
             dtype='datetime64[ns]', freq='D')
```

```
# Values
df.values
```

```
array([[ 1.17503197, -2.2455333 ,  1.19639255, -1.89623003],
       [ 0.21165485, -0.93104948,  0.33932534, -0.99199535],
       [ 1.54112107,  0.70958415,  1.32130367,  0.71557556],
       [-0.18062483, -1.33214427, -0.50359153, -0.45864285],
       [ 1.02492346, -1.35643648, -2.66123573,  0.76561735],
       [-0.20947413, -0.73914306,  0.07642315,  2.34669621]])
```

```
# Statistical summary
df.describe()
```

```
              A         B         C         D
count  6.000000  6.000000  6.000000  6.000000
mean   0.593772 -0.982454 -0.038564  0.080170
std    0.749951  0.977993  1.457741  1.507099
min   -0.209474 -2.245533 -2.661236 -1.896230
25%   -0.082555 -1.350363 -0.358588 -0.858657
50%    0.618289 -1.131597  0.207874  0.128466
75%    1.137505 -0.787120  0.982126  0.753107
max    1.541121  0.709584  1.321304  2.346696
```

```
# Transpose
df.T
```

```
    2013-01-01  2013-01-02  2013-01-03  2013-01-04  2013-01-05 ␣
↪2013-01-06
A     1.175032    0.211655    1.541121   -0.180625    1.024923 ␣
↪  -0.209474
B    -2.245533   -0.931049    0.709584   -1.332144   -1.356436 ␣
↪  -0.739143
C     1.196393    0.339325    1.321304   -0.503592   -2.661236 ␣
↪   0.076423
D    -1.896230   -0.991995    0.715576   -0.458643    0.765617 ␣
↪   2.346696
```

```
# Sort by an axis
df.sort_index(axis=1, ascending=False)
```

```
                    D            C            B            A
2013-01-01  -1.896230     1.196393    -2.245533     1.175032
2013-01-02  -0.991995     0.339325    -0.931049     0.211655
2013-01-03   0.715576     1.321304     0.709584     1.541121
2013-01-04  -0.458643    -0.503592    -1.332144    -0.180625
2013-01-05   0.765617    -2.661236    -1.356436     1.024923
2013-01-06   2.346696     0.076423    -0.739143    -0.209474
```

```
# Sort by values
df.sort_values(by="B")
```

```
                    A            B            C            D
2013-01-01   1.175032    -2.245533     1.196393    -1.896230
2013-01-05   1.024923    -1.356436    -2.661236     0.765617
2013-01-04  -0.180625    -1.332144    -0.503592    -0.458643
2013-01-02   0.211655    -0.931049     0.339325    -0.991995
2013-01-06  -0.209474    -0.739143     0.076423     2.346696
2013-01-03   1.541121     0.709584     1.321304     0.715576
```

Selections

One of the first steps in data analysis is simply to filter your data and get slices you're most interested in. Pandas has numerous approaches to quickly get only what you want.

- Select a single column by addressing the dataframe as you would a dictionary, with df[column] or, if the column name is a single word, with df.column. This returns a series.
- Select a slice in the way you would a Python list, with df[], e.g. df[:3], or by slicing the indices, df["20130102":"20130104"].
- Use .loc to select by specific labels, such as:
  - Get a cross-section based on a label, with e.g. df.loc[index[0]].
  - Get on multi-axis by a label, with df.loc[:, ["A", "B"]] where the first : indicates the slice of rows, and the second list ["A", "B"] indicates the list of columns.
- As you would with Numpy, you can get a boolean-based selection, with e.g. df[df.A > num].

There are a *lot* more ways to filter and access data, as well as methods to set data in your dataframes, but this will be enough for now.

```
# By column
df.A
```

```
2013-01-01     1.175032
2013-01-02     0.211655
2013-01-03     1.541121
2013-01-04    -0.180625
2013-01-05     1.024923
2013-01-06    -0.209474
Freq: D, Name: A, dtype: float64
```

```
# By slice
df["20130102":"20130104"]
```

```
                    A          B          C          D
2013-01-02   0.211655  -0.931049   0.339325  -0.991995
2013-01-03   1.541121   0.709584   1.321304   0.715576
2013-01-04  -0.180625  -1.332144  -0.503592  -0.458643
```

```
# Cross-section
df.loc[dates[0]]
```

```
A     1.175032
B    -2.245533
C     1.196393
D    -1.896230
Name: 2013-01-01 00:00:00, dtype: float64
```

```
# Multi-axis
df.loc[:, ["A", "B"]]
```

```
                    A          B
2013-01-01   1.175032  -2.245533
2013-01-02   0.211655  -0.931049
2013-01-03   1.541121   0.709584
2013-01-04  -0.180625  -1.332144
2013-01-05   1.024923  -1.356436
2013-01-06  -0.209474  -0.739143
```

```
# Boolean indexing
df[df.A > 0]
```

```
                    A          B          C          D
2013-01-01   1.175032  -2.245533   1.196393  -1.896230
2013-01-02   0.211655  -0.931049   0.339325  -0.991995
2013-01-03   1.541121   0.709584   1.321304   0.715576
2013-01-05   1.024923  -1.356436  -2.661236   0.765617
```

7.2.3   *Matplotlib*

In this last section, you get to meet *Matplotlib*, a fairly ubiquitous and powerful Python plotting library. Jupyter Notebook has some "magic" we can use in the line %matplotlib inline, which permits us to draw charts directly in this notebook.

Matplotlib, Numpy, and Pandas form the three most important and ubiquitous tools in data analysis.

Note that this is the merest slither of an introduction to what you can do with these libraries.

```python
import matplotlib.pyplot as plt
# This bit of magic code will allow your Matplotlib plots to
↪be shown directly in your Jupyter Notebook.
%matplotlib inline

# Produce a random time series
ts = pd.Series(np.random.randn(1000), index=pd.date_range('1/1/
↪2000', periods=1000))

# Get the cumulative sum of the random numbers generated to
↪mimic a historic data series
ts = ts.cumsum()

# And magically plot
ts.plot()
```

```
<matplotlib.axes._subplots.AxesSubplot at 0x1f537de6860>
```

```python
# And do the same thing with a dataframe
df = pd.DataFrame(np.random.randn(1000, 4), index=ts.index,
                  columns=['A', 'B', 'C', 'D'])

df = df.cumsum()

# And plot, this time creating a figure and adding a plot and
↪legend to it
plt.figure()
df.plot()
plt.legend(loc='best')
```

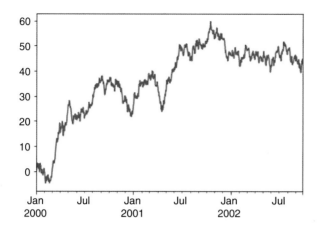

```
<matplotlib.legend.Legend at 0x1f538166c50>
```

```
<Figure size 432x288 with 0 Axes>
```

7.3 Sample Code

```
print("hello")

a=1

print(a)

b="abc"

print(b)

print("Please input your name")
x = input()
print("Your name is : ", x)
```

```
hello
1
abc
Please input your name
Your name is :  tom
```

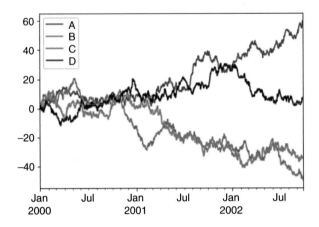

```
if a == 1:
    print("a is equal to 1")
else:
    print("a is not equal to 1")

for i in range (2,6):
    print(i)

# compute area
print("please input the radius : ")
x = float(input())
area = 3.142 * (x ** 2)
print("the area is ", area)

def cal_area(r):
    a = 3.142 * (r ** 2)
    return a

print("please input the radius : ")
radius = float(input())
area = cal_area(radius)
print("the area is ", area)
```

```
a is equal to 1
2
3
4
5
please input the radius :
the area is  28.278
please input the radius :
the area is  12.568
```

```
# Class
class Basic:
    x=3

y = Basic()
print(y.x)

class Computation:
    def area(self):
        return 3.142*self.radius**2
    def parameter(self):
        return 3.142*self.radius*2
    def __init__(self, radius):
        self.radius = radius

a = Computation(3)
print("Area is : ", a.area())

print("Parameter is : ", a.parameter())
```

```
3
Area is :   28.278
Parameter is :   18.852
```

```
def adder(*num):
    sum = 0

    for n in num:
        sum = sum + n

    print("Sum:",sum)

adder(3,5)
adder(4,5,6,7)
adder(1,2,3,5,6)

def intro(**data):
    print("Data type of argument: ",type(data), "\n")

    for key, value in data.items():
        print("{} is {}".format(key,value))

intro(Firstname="Sita", Lastname="Sharma", Age=22,
→Phone=1234567890)
intro(Firstname="John", Lastname="Wood", Email=
→"johnwood@nomail.com", Country="Wakanda", Age=25,
→Phone=9876543210)
```

```
Sum:  8
Sum:  22
Sum:  17
Data type of argument:   <class 'dict'>

Firstname is Sita
Lastname is Sharma
Age is 22
Phone is 1234567890
Data type of argument:   <class 'dict'>

Firstname is John
Lastname is Wood
Email is johnwood@nomail.com
Country is Wakanda
Age is 25
Phone is 9876543210
```

And that's it for this quick introduction to Python and its use in data analysis.

Part II
Artificial Intelligence Basics

Chapter 8
Introduction to Artificial Intelligence

Abstract Humans can accomplish tasks that scientists are still trying to fathom, and such tasks are hard to write algorithms for. Artificial Intelligence programs are thus written in a way that allows these algorithms to learn from data. This makes data quality crucial to the performance of the algorithm. Data exploration and investigation are a must for Artificial Intelligence developers to identify appropriate charts, present data to visualize its core characteristics, and tell stories observed.

Learning outcomes:

- Investigate and manipulate data to learn its metadata, shape, and robustness.
- Identify an appropriate chart and present data to illustrate its core characteristics.
- Aggregate and present data-driven analysis using NumPy, Pandas, and Matplotlib.

In the 1950s, Artificial Intelligence was viewed as a futuristic, theoretical part of computer science. Now, due to increases in computing capacity and extensive research into algorithms, Artificial Intelligence is now a viable reality. So much so that many of the products we use every day have some variations of Artificial Intelligence built into them.

Artificial Intelligence is a program for a machine in order to execute tasks that a human could do. Writing algorithms for tasks that humans can do can be very challenging. The human mind can do many things that scientists could not understand, much less approximate. For example, how would we write algorithms for these tasks?

1. A song comes on the radio, and most listeners of music can quickly identify the genre, maybe the artist, and probably the song.
2. An art critic sees a painting he has never seen before, yet he could most likely identify the era, the medium, and probably the artist.
3. A baby can recognize her mom's face at a very early age.

The simple answer is that you cannot write algorithms for these. Algorithms use mathematics. Humans who accomplish these tasks could not explain mathematically how they drew these conclusions. They were able to achieve these results because they learned to do these things over time. Artificial Intelligence was designed to simulate human learning on a computer.

T. T. Teoh, Z. Rong, *Artificial Intelligence with Python*,
Machine Learning: Foundations, Methodologies, and Applications,
https://doi.org/10.1007/978-981-16-8615-3_8

In fact, during the early stages of Artificial Intelligence research, the researchers began with developing algorithms to try to approximate human intuition. This code could be viewed as a huge if/else statement which produces the answer. This turned out to be an incredibly inefficient approach due to the complexity of the human mind. The rules are very rigid and are most likely to become obsolete as circumstances change over time.

Instead of trying to program a machine to act as a brain, why do not we just feed it a bunch of data so that it can figure out the best algorithm on its own? That is where machine learning algorithms would come into play. Machine learning algorithms are the engine of practically every Artificial Intelligence system.

Machine learning is what enables smart systems to get smarter. These algorithms are designed to equip Artificial Intelligence with the power to self-educate and improve its own accuracy over time, learning from the data it is steadily taking in. This means the Artificial Intelligence is always adjusting to interactions between data points, providing living, breathing data analysis as the data quality changes.

And because machine learning is an iterative process, the data quality, particularly early on, is crucial to performance. AI that gets trained on datasets with anomalies or incorrectly tagged information will lead to false positives and less effective machine learning.

Therefore, the quality of the data used must be good in order to create good artificial intelligence programs. Both Artificial Intelligence engineers and data scientists have to be very data-savvy for data processing. Hence, the skills required often overlap.

There are a number of tools used by Artificial Intelligence engineers and data scientists to understand and analyze data. We will get to those, but one of the fundamentals is simply exploring a new dataset.

8.1 Data Exploration

Usually, in data courses, you are presented with a nice clean dataset and run some algorithms on it and get some answers. That is not helpful to you. Except for data you collect, you are unlikely to know the shape and contents of a dataset you import from others, no matter how good their research.

Especially for large datasets, it can be difficult to know how many unique terms you may be working with and how they relate to each other.

Open your own Jupyter Notebook and follow along with the code:

```
# Comments to code are not executed and are flagged with this '
↪#' symbol.
# First we'll import the pandas library.
# We use 'as' so that we can reference it as 'pd', which is␣
↪shorter to type.
import pandas as pd
```

(continues on next page)

(continued from previous page)

```python
# In Python, we can declare our variables by simply naming
↪them, like below
# 'data_url' is a variable name and the url is the text
↪reference we're assigning to it
data_url = "https://docs.google.com/spreadsheets/d/
↪1P0ob0sfz3xqG8u_dxT98YcVTMwzPSnya_qx6MbX-_Z8/pub?gid=0&
↪single=true&output=csv"

# We import our data as a 'dataframe' using this simple
↪instruction.
# How did I know it was a CSV file? If you look at the end of
↪the urls (above), you'll see 'output=csv'.
# A variable in Python can be anything. Here our variable is a
↪Pandas dataframe type.
data = pd.read_csv(data_url)

# Let us see what that looks like (I limit the number of rows
↪printed by using '[:10]',
# and Python is '0' indexed, meaning the first term starts at
↪'0'):
data[:10]
```

```
        Date         Governorate    Cases  Deaths  CFR (%)  \
0  2018-02-18               Amran   103965     176     0.17
1  2018-02-18           Al Mahwit    62887     151     0.24
2  2018-02-18          Al Dhale'e    47136      81     0.17
3  2018-02-18              Hajjah   121287     422     0.35
4  2018-02-18              Sana'a    76250     123     0.16
5  2018-02-18              Dhamar   103214     161     0.16
6  2018-02-18               Abyan    28243      35     0.12
7  2018-02-18          Al Hudaydah  155908     282     0.18
8  2018-02-18            Al Bayda    30568      36     0.12
9  2018-02-18      Amanat Al Asimah 103184      71     0.07

   Attack Rate (per 1000)  COD Gov English COD Gov Arabic  COD Gov Pcode
0                  89.582            Amran        عمران            29.0
1                  86.122        Al Mahwit      المحويت           27.0
2                  64.438       Al Dhale'e       الضالع           30.0
3                  52.060           Hajjah         حجة            17.0
4                  51.859           Sana'a       صنعاء            23.0
5                  51.292           Dhamar         ذمار           20.0
6                  49.477            Abyan         أبين           12.0
7                  48.147        Al Hudaydah     الحديدة          18.0
8                  40.253         Al Bayda       البيضاء          14.0
9                  36.489   Amanat Al Asimah  أمانة العاصمة        13.0
```

These are the first ten rows of the Pandas dataframe. You can think of a dataframe as being like a database table allowing you to do bulk operations, or searches and filters, on the overall data.

The top bolded row of the dataframe contains the terms which describe the data in each column. Not all of those terms will be familiar, and—even when familiar—the units may not be obvious. These *headers* are another form of metadata.

The data about the overall dataset is called *descriptive metadata*. Now we need information about the data within each dataset. That is called **structural metadata**, a grammar describing the structure and definitions of the data in a table.

Sometimes the data you are working with has no further information and you need to experiment with similar data to assess what the terms mean, or what unit is being used, or to gap-fill missing data. Sometimes there is someone to ask. Sometimes you get a structural metadata definition to work with.

This process, of researching a dataset, of exploration and tremendous frustration, is known as *munging* or *data wrangling*.

In this case, the publisher has helpfully provided another table containing the definitions for the structural metadata.

```
# First we set the url for the metadata table
metadata_url = "https://docs.google.com/spreadsheets/d/
↪1P0ob0sfz3xqG8u_dxT98YcVTMwzPSnya_qx6MbX-_Z8/pub?
↪gid=771626114&single=true&output=csv"
# Import it from CSV
metadata = pd.read_csv(metadata_url)
# Show the metadata:
metadata
```

```
                           Column                                     ␣
 ↪     Description
0                            Date                  Date when the figures␣
↪were reported.
1                    Governorate   The Governorate name as reported␣
↪in the WHO ep...
2                           Cases   Number of cases recorded in the␣
↪governorate si...
3                          Deaths   Number of deaths recorded in the␣
↪governorate s...
4                         CFR (%)   The case fatality rate in␣
↪governorate since 27...
5   Attack Rate (per 1000)   The attack rate per 1,000 of the␣
↪population in...
6              COD Gov English   The English name for the␣
↪governorate according...
7               COD Gov Arabic   The Arabic name for the␣
↪governorate according ...
8               COD Gov Pcode   The PCODE name for the governorate␣
↪according t...
9                 Bulletin Type   The type of bulletin from which␣
↪the data was e...
10                Bulletin URL   The URL of the bulletin from which␣
↪the data wa...
```

The column widths are too narrow to read the full text. There are two ways we can widen them. The first is to adjust the output style of the dataframe. The second is to pull out the text from each cell and iterate through a list. The first is easier (one line), but the second is an opportunity to demonstrate how to work with dataframes.

We can explore each of these metadata terms, but rows 2 to 5 would appear the most relevant.

```
# First, the one-line solution
metadata[2:6].style.set_properties(subset=['Description'], **{
 ↪'width': '400px', 'text-align': 'left'})
```

```
<pandas.io.formats.style.Styler at 0x1a22e3fbef0>
```

The second approach is two lines and requires some new coding skills. We address an individual cell from a specific dataframe column as follows:

```
dataframe.column_name[row_number]
```

We have four terms and it would be tedious to type out each term we are interested in this way, so we will use a loop. Python uses whitespace indentation to structure its code.

```
for variable in list:
    print(variable)
```

This will loop through the list of variables you have, giving the name `variable` to each item. Everything indented (using either a tab or four spaces to indent) will be executed in order in the loop. In this case, the loop prints the variable.

We are also going to use two other code terms:

- `'{}{}'.format(var1, var2)`—used to add variables to text; `{}` braces will be replaced in the order the variables are provided
- `range`—a way to create a numerical list (e.g., `range(2,6)` creates a list of integers like this `[2,3,4,5]`)

```
for i in range(2, 6):
    print('{} - {}'.format(i, metadata.Description[i]))
```

```
2 - Number of cases recorded in the governorate since 27 April␣
↪2017.
3 - Number of deaths recorded in the governorate since 27␣
↪April 2017.
4 - The case fatality rate in governorate since 27 April 2017.
5 - The attack rate per 1,000 of the population in the␣
↪governorate since 27 April 2017.
```

Unless you work in epidemiology, "attack rate" may still be unfamiliar. The US Centers for Disease Control and Prevention has a self-study course which covers the principles of epidemiology and contains this definition: "In the outbreak setting, the term attack rate is often used as a synonym for risk. It is the risk of getting the disease during a specified period, such as the duration of an outbreak."

An "Attack rate (per 1000)" implies the rate of new infections per 1,000 people in a particular population.

There are two more things to find out: how many governorates are there in Yemen, and over what period do we have data?

```
# Get the unique governorates from the 'Governorate' column:
# Note the way we address the column and call for 'unique()'
governorates = data.Governorate.unique()
print("Number of Governorates: {}".format(len(governorates)))
print(governorates)
```

```
Number of Governorates: 26
['Amran' 'Al Mahwit' "Al Dhale'e" 'Hajjah' "Sana'a" 'Dhamar'
 ↪'Abyan'
 'Al Hudaydah' 'Al Bayda' 'Amanat Al Asimah' 'Raymah' 'Al Jawf
 ↪' 'Lahj'
 'Aden' 'Ibb' 'Taizz' 'Marib' "Sa'ada" 'Al Maharah' 'Shabwah'
 ↪'Moklla'
 "Say'on" 'Al-Hudaydah' 'Al_Jawf' "Ma'areb" 'AL Mahrah']
```

```
# We can do the same for the dates, but we also want to know␣
↪the start and end
# Note the alternative way to address a dataframe column
date_list = data["Date"].unique()
print("Starting on {}, ending on {}; with {} periods.".
↪format(min(date_list), max(date_list), len(date_list)))
```

```
Starting on 2017-05-22, ending on 2018-02-18; with 136 periods.
```

We can now summarize what we have learned: data covering a daily update of cholera infection and fatality rates for 131 days, starting on 22 May till 14 January 2018 for the 26 governorates in Yemen.

Mostly this confirms what was in the description on HDX, but we also have some updates and additional data to consider.

Before we go any further, it is helpful to check that the data presented are in the format we expect. Are those integers and floats defined that way, or are they being interpreted as text (because, for example, someone left commas in the data)?

```
# This will give us a quick summary of the data, including␣
↪value types and the number or rows with valid data
data.info()
```

```
<class 'pandas.core.frame.DataFrame'>
RangeIndex: 2914 entries, 0 to 2913
Data columns (total 9 columns):
Date                    2914 non-null object
Governorate             2914 non-null object
Cases                   2914 non-null object
Deaths                  2914 non-null int64
CFR (%)                 2914 non-null float64
Attack Rate (per 1000)  2914 non-null float64
```

(continues on next page)

(continued from previous page)

```
COD Gov English              2713 non-null  object
COD Gov Arabic               2713 non-null  object
COD Gov Pcode                2713 non-null  float64
dtypes: float64(3), int64(1), object(5)
memory usage: 205.0+ KB
```

Immediately we notice a problem. Our "Dates" are not of a date-type, and "Cases" are not integers. This will cause us problems as we get deeper into our analysis. Thankfully, conversion is quick and easy.

Pandas will attempt to figure out what type of object each value is and assign it appropriately. If it cannot figure it out, then the default is to convert it to a text string. There are a number of ways in which numbers and dates can be formatted confusingly, and the most common for integers is when commas are included.

Ordinarily—especially if we intend to download each update of these data and use the series regularly—we can specify transformations on load. We did not know about these problems when we started, so we will fix it directly now.

These sorts of transformations—converting dates, integers, and floats—are common requirements and it is useful to get a grasp of them at the beginning. In later modules, we will consider more complex transformations and gap-filling.

```
# Removing commas for an entire column and converting to␣
↪integers
data["Cases"] = [int(x.replace(",","")) for x in data["Cases"]]
# And converting to date is even simpler
data["Date"] = pd.to_datetime(data["Date"])
# And let's check the overview again
data.info()
```

```
<class 'pandas.core.frame.DataFrame'>
RangeIndex: 2914 entries, 0 to 2913
Data columns (total 9 columns):
Date                     2914 non-null  datetime64[ns]
Governorate              2914 non-null  object
Cases                    2914 non-null  int64
Deaths                   2914 non-null  int64
CFR (%)                  2914 non-null  float64
Attack Rate (per 1000)   2914 non-null  float64
COD Gov English          2713 non-null  object
COD Gov Arabic           2713 non-null  object
COD Gov Pcode            2713 non-null  float64
dtypes: datetime64[ns](1), float64(3), int64(2), object(3)
memory usage: 205.0+ KB
```

The code used to transform "Cases" (the bit to the right of = and between the []) is called a *list comprehension*. These are very efficient, taking little time to execute.

The time it takes code to run is not a major concern right now, with only 2,803 rows, but it becomes a major factor once we work with larger datasets and is something addressed in later modules.

Our data are a time series and our analysis will focus on attempting to understand what is happening and where. We are continuing to explore the shape of it and assessing how we can best present the human story carried by that data.

We know that the cholera epidemic is getting worse, since more governorates were added in since the time series began. To get a rough sense of how the disease and humanitarian response has progressed, we will limit our table only to the columns we are interested in and create two slices at the start and end of the series.

```
# First, we limit our original data only to the columns we
↪will use,
# and we sort the table according to the attack rate:
data_slice = data[["Date", "Governorate", "Cases", "Deaths",
↪"CFR (%)", "Attack Rate (per 1000)"]
                    ].sort_values("Attack Rate (per 1000)",
↪ascending=False)
# Now we create our two slices, and set the index to
↪Governorate
ds_start = data_slice.loc[data_slice.Date == "2017-05-22"].set_
↪index("Governorate")
ds_end = data_slice.loc[data_slice.Date == "2018-01-14"].set_
↪index("Governorate")
# And print
print(ds_start)
print(ds_end)
```

	Date	Cases	Deaths	CFR (%)	Attack Rate (per 1000)
Governorate					
Al Mahwit	2017-05-22	2486	34	1.4	3.27
Sana'a	2017-05-22	3815	39	1.0	3.05
Amanat Al Asimah	2017-05-22	9216	33	0.4	2.79
Amran	2017-05-22	3743	45	1.2	2.45
Hajjah	2017-05-22	4664	42	0.9	2.10
Al Bayda	2017-05-22	1498	6	0.4	1.95
Al Dhale'e	2017-05-22	1401	8	0.6	1.86
Abyan	2017-05-22	1068	10	0.9	1.75
Raymah	2017-05-22	549	4	0.7	0.87
Dhamar	2017-05-22	1617	33	2.0	0.76
Taizz	2017-05-22	1791	23	1.3	0.59
Aden	2017-05-22	489	12	2.5	0.51

(continues on next page)

(continued from previous page)

	Date	Cases	Deaths	CFR (%)	Attack↵Rate (per 1000)
Ibb	2017-05-22	1378	37	2.7	0.45
Al-Hudaydah	2017-05-22	1397	32	2.3	0.42
Al_Jawf	2017-05-22	189	3	1.6	0.29
Lahj	2017-05-22	168	0	0.0	0.16
Ma'areb	2017-05-22	2	0	0.0	0.01

	Date	Cases	Deaths	CFR (%)	Attack↵Rate (per 1000)
Governorate					
Amran	2018-01-14	102231	175	0.17	88.088
Al Mahwit	2018-01-14	61097	149	0.24	83.671
Al Dhale'e	2018-01-14	47132	81	0.17	64.432
Hajjah	2018-01-14	118468	420	0.35	50.850
Sana'a	2018-01-14	74103	123	0.17	50.398
Dhamar	2018-01-14	99766	160	0.16	49.579
Abyan	2018-01-14	28241	35	0.12	49.473
Al Hudaydah	2018-01-14	150965	280	0.19	46.620
Al Bayda	2018-01-14	28730	34	0.12	37.833
Amanat Al Asimah	2018-01-14	99452	70	0.07	35.169
Al Jawf	2018-01-14	15827	22	0.14	27.355
Raymah	2018-01-14	16403	120	0.73	27.033
Lahj	2018-01-14	24341	22	0.09	24.127
Aden	2018-01-14	20868	62	0.30	22.609
Ibb	2018-01-14	64536	286	0.44	21.824
Taizz	2018-01-14	62371	187	0.30	20.802
Marib	2018-01-14	7285	7	0.10	20.318
Sa'ada	2018-01-14	10703	5	0.05	11.825
Al Maharah	2018-01-14	1168	1	0.09	7.866

(continues on next page)

(continued from previous page)

Shabwah		2018-01-14	1399	3	0.21	␣
↪	2.315					
Moklla		2018-01-14	568	2	0.35	␣
↪	1.417					
Say'on		2018-01-14	22	0	0.00	␣
↪	0.100					

There is a great deal of data to process here, but the most important is that the attack rate has risen exponentially, and cholera has spread to more areas.

However, there are also a few errors in the data. Note that *Al Jawf* appears twice (as "Al Jawf" and as "Al_Jawf"). It is essential to remember that computers are morons. They can only do exactly what you tell them to do. Different spellings, or even different capitalizations, of words are different words.

You may have hoped that the data munging part was complete, but we need to fix this. We should also account for the introduction of "Moklla" and "Say'on," which are two districts in the governorate of "Hadramaut" so that we do only have a list of governorates (and you may have picked this up if you'd read through the comments in the metadata earlier).

We can now filter our dataframe by the groups of governorates we need to correct. This introduces a few new concepts in Python. The first of these is that of a *function*. This is similar to the libraries we have been using, such as Pandas. A function encapsulates some code into a reusable object so that we do not need to repeat ourselves and can call it whenever we want.

```python
def fix_governorates(data, fix_govs):
    """
    This is our function _fix_governorates_; note that we must␣
↪pass it
    two variables:
        - data: the dataframe we want to fix;
        - fix_govs : a dictionary of the governorates we need␣
↪to correct.

    The function will do the following:
        For a given dataframe, date list, and dictionary of␣
↪Governorates
        loop through the keys in the dictionary and combine␣
↪the list
        of associated governorates into a new dataframe.
        Return a new, corrected, dataframe.
    """
    # Create an empty list for each of the new dataframes we
↪'ll create
    new_frames = []
    # And an empty list for all the governorates we'll need to␣
↪remove later
    remove = []
    # Create our list of dates
```

(continues on next page)

(continued from previous page)

```python
    date_list = data["Date"].unique()
    # Loop through each of the governorates we need to fix
    for key in fix_govs.keys():
        # Create a filtered dataframe containing only the
↪governorates to fix
        ds = data.loc[data_slice.Governorate.isin(fix_
↪govs[key])]
        # New entries for the new dataframe
        new_rows = {"Date": [],
                    "Cases": [],
                    "Deaths": [],
                    "CFR (%)": [],
                    "Attack Rate (per 1000)": []
                   }
        # Divisor for averages (i.e., there could be more than
↪2 govs to fix)
        num = len(fix_govs[key])
        # Add the governorate values to the remove list
        remove.extend(fix_govs[key])
        # For each date, generate new values
        for d in date_list:
            # Data in the dataframe is stored as a Timestamp
↪value
            r = ds[ds["Date"] == pd.Timestamp(d)]
            new_rows["Date"].append(pd.Timestamp(d))
            new_rows["Cases"].append(r.Cases.sum())
            new_rows["Deaths"].append(r.Deaths.sum())
            new_rows["CFR (%)"].append(r["CFR (%)"].sum()/num)
            new_rows["Attack Rate (per 1000)"].append(r[
↪"Attack Rate (per 1000)"].sum()/num)
        # Create a new dataframe from the combined data
        new_rows = pd.DataFrame(new_rows)
        # And assign the values to the key governorate
        new_rows["Governorate"] = key
        # Add the new dataframe to our list of new frames
        new_frames.append(new_rows)
    # Get an inverse filtered dataframe from what we had before
    ds = data_slice.loc[~data_slice.Governorate.isin(remove)]
    new_frames.append(ds)
    # Return a new concatenated dataframe with all our
↪corrected data
    return pd.concat(new_frames, ignore_index=True)
```

Now we can run our function on our data and reproduce the two tables from before.

```python
fix = {"Hadramaut": ["Moklla","Say'on"],
       "Al Hudaydah": ["Al Hudaydah", "Al-Hudaydah"],
       "Al Jawf": ["Al Jawf", "Al_Jawf"],
       "Al Maharah": ["Al Maharah", "AL Mahrah"],
       "Marib": ["Marib", "Ma'areb"]
```

(continues on next page)

(continued from previous page)

```
    }
# Using %time to see how long this takes
%time data_slice = fix_governorates(data_slice, fix).sort_
↪values("Attack Rate (per 1000)", ascending=False)
# Now we recreate our two slices, and set the index to␣
↪Governorate
ds_start = data_slice.loc[data_slice.Date == "2017-05-22"].set_
↪index("Governorate")
ds_end = data_slice.loc[data_slice.Date == "2018-01-14"].set_
↪index("Governorate")
# And print
print(ds_start)
print(ds_end)
```

```
Wall time: 750 ms
                     Attack Rate (per 1000)   CFR (%)   Cases          ␣
↪Date   Deaths
Governorate
Al Mahwit                              3.270      1.40    2486 2017-
↪05-22       34
Sana'a                                 3.050      1.00    3815 2017-
↪05-22       39
Amanat Al Asimah                       2.790      0.40    9216 2017-
↪05-22       33
Amran                                  2.450      1.20    3743 2017-
↪05-22       45
Hajjah                                 2.100      0.90    4664 2017-
↪05-22       42
Al Bayda                               1.950      0.40    1498 2017-
↪05-22        6
Al Dhale'e                             1.860      0.60    1401 2017-
↪05-22        8
Abyan                                  1.750      0.90    1068 2017-
↪05-22       10
Raymah                                 0.870      0.70     549 2017-
↪05-22        4
Dhamar                                 0.760      2.00    1617 2017-
↪05-22       33
Taizz                                  0.590      1.30    1791 2017-
↪05-22       23
Aden                                   0.510      2.50     489 2017-
↪05-22       12
Ibb                                    0.450      2.70    1378 2017-
↪05-22       37
Al Hudaydah                            0.210      1.15    1397 2017-
↪05-22       32
Lahj                                   0.160      0.00     168 2017-
↪05-22        0
Al Jawf                                0.145      0.80     189 2017-
↪05-22        3
Marib                                  0.005      0.00       2 2017-
↪05-22        0
```

(continues on next page)

(continued from previous page)

Governorate	Attack Rate (per 1000)	CFR (%)	Cases	Date	Deaths
Hadramaut	0.000	0.00	0	2017-05-22	0
Al Maharah	0.000	0.00	0	2017-05-22	0
Amran	88.0880	0.170	102231	2018-01-14	175
Al Mahwit	83.6710	0.240	61097	2018-01-14	149
Al Dhale'e	64.4320	0.170	47132	2018-01-14	81
Hajjah	50.8500	0.350	118468	2018-01-14	420
Sana'a	50.3980	0.170	74103	2018-01-14	123
Dhamar	49.5790	0.160	99766	2018-01-14	160
Abyan	49.4730	0.120	28241	2018-01-14	35
Al Bayda	37.8330	0.120	28730	2018-01-14	34
Amanat Al Asimah	35.1690	0.070	99452	2018-01-14	70
Raymah	27.0330	0.730	16403	2018-01-14	120
Lahj	24.1270	0.090	24341	2018-01-14	22
Al Hudaydah	23.3100	0.095	150965	2018-01-14	280
Aden	22.6090	0.300	20868	2018-01-14	62
Ibb	21.8240	0.440	64536	2018-01-14	286
Taizz	20.8020	0.300	62371	2018-01-14	187
Al Jawf	13.6775	0.070	15827	2018-01-14	22
Sa'ada	11.8250	0.050	10703	2018-01-14	5
Marib	10.1590	0.050	7285	2018-01-14	7
Al Maharah	3.9330	0.045	1168	2018-01-14	1
Shabwah	2.3150	0.210	1399	2018-01-14	3
Hadramaut	0.7585	0.175	590	2018-01-14	2

```
C:\Users\turuk\Anaconda3\envs\calabar\lib\site-packages\
↪ipykernel_launcher.py:54: FutureWarning: Sorting because non-
↪concatenation axis is not aligned. A future version
of pandas will change to not sort by default.

To accept the future behavior, pass 'sort=False'.

To retain the current behavior and silence the warning, pass
↪'sort=True'.
```

We can also create a line chart to see how the number of cases has progressed over time. This will be our first use of *Matplotlib*, a fairly ubiquitous and powerful Python plotting library. Jupyter Notebook has some "magic" we can use in the line %matplotlib inline, which permits us to draw charts directly in this notebook.

```
# Matplotlib for additional customization
from matplotlib import pyplot as plt
%matplotlib inline

# First we create a pivot table of the data we wish to plot.␣
↪Here only the "Cases", although you
# should experiment with the other columns as well.
drawing = pd.pivot_table(data_slice, values="Cases", index=[
↪"Date"], columns=["Governorate"])
# Then we set a plot figure size and draw
drawing.plot(figsize=(20,15), grid=False)
```

```
<matplotlib.axes._subplots.AxesSubplot at 0x1a22efee4e0>
```

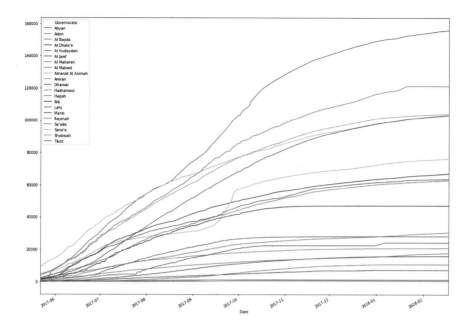

These are not glamorous charts or tables. This last is what I call a *spaghetti chart* because of the tangle of lines that make it difficult to track what is happening.

However, they are useful methods for investigating what the data tell us and contextualizing it against the events behind the data.

Perhaps, given where we are, you feel some confidence that you could begin to piece together a story of what is happening in the Yemen cholera epidemic?

8.2 Problems with Data

We will be exploring the data and the trouble with accuracy.

Sitting at your computer in comfortable surroundings—whether in a quiet office or the clatter and warmth of your favorite coffee shop—it is tempting to place confidence in a neat table of numbers and descriptions. You may have a sense that data are, in some reassuring way, *truthy*.

They are not.

All data are a reflection of the time when they were collected, the methodology that produced it, and the care with which that methodology was implemented. It is a sample of a moment in time and it is inherently imperfect.

Medical data, produced by interviewing patient volunteers, is reliant on self-reported experiences and people—even when they are trying to be honest and reporting on something uncontroversial—have imperfect memories. Blood or tissue samples depend on the consistency with which those samples were acquired, and the chain that stretches from patient, to clinic, to courier, to laboratory, and to data analyst. Anything can go wrong, from spillage to spoilage to contamination to overheating or freezing.

Even data generated autonomously via sensors or computational sampling is based on what a human thought was important to measure and implemented by people who had to interpret instructions on what to collect and apply it to the tools at hand. Sensors can be in the wrong place, pointing in the wrong direction, miscalibrated, or based on faulty assumptions from the start.

Data carry the bias of the people who constructed the research and the hopes of those who wish to learn from it.

Data are inherently uncertain and any analysis must be absolutely cognizant of this. It is the reason we start with ethics. We must, from the outset, be truthful to ourselves.

In future lessons, we will consider methods of assessing the uncertainty in our data and how much confidence we can have. For this lesson, we will develop a theoretical understanding of the uncertainty and which data we can use to tell a story about events happening in Yemen.

In the space of six months (from May to November 2017), Yemen went from 35,000 cholera cases to almost 1 million. Deaths now exceed 2000 people per month and the attack rate per 1000 has gone from an average of 1 to 30. This reads like an out-of-control disaster.

At the same time, however, the fatality rate has dropped from 1% to 0.2%.

Grounds for optimism, then? Somehow medical staff are getting on top of the illness even as infection spreads?

Consider how these data are collected. Consider the environment in which it is being collected.

Background: reading on what is happening in Yemen (December 2017):

- Yemen: Coalition Blockade Imperils Civilians—Human Rights Watch, 7 December 2017
- What is happening in Yemen and how are Saudi Arabia's airstrikes affecting civilians—Paul Torpey, Pablo Gutiérrez, Glenn Swann and Cath Levett, The Guardian, 16 September 2016
- Saudi "should be blacklisted" over Yemen hospital attacks—BBC, 20 April 2017
- Process Lessons Learned in Yemen's National Dialogue—Erica Gaston, USIP, February 2014. According to UNICEF, as of November 2017, "More than 20 million people, including over 11 million children, are in need of urgent humanitarian assistance. At least 14.8 million are without basic healthcare and an outbreak of cholera has resulted in more than 900,000 suspected cases."

Cholera incidence data are being collected in an active war zone where genocide and human rights violations committed daily. Hospital staff are stretched thin, and many have been killed. Islamic religious law requires a body to be buried as soon as possible, and this is even more important in a conflict zone to limit further spread of disease.

The likelihood is that medical staff are overwhelmed and that the living and ill must take precedence over the dead. They see as many people as they can, and it is a testament to their dedication and professionalism that these data continue to reach the WHO and UNICEF.

There are human beings behind these data. They have suffered greatly to bring it to you.

In other words, all we can be certain of is that the Cases and Deaths are the minimum likely and that attack and death rates are probably extremely inaccurate. The undercount in deaths may lead to a false sense that the death rate is falling relative to infection, but one should not count on this.

Despite these caveats, humanitarian organizations must use these data to prepare their relief response. Food, medication, and aid workers must be readied for the moment when fighting drops sufficiently to get to Yemen. Journalists hope to stir public opinion in donor nations (and those outside nations active in the conflict), using these data to explain what is happening.

The story we are working on must accept that the infection rate is the only data that carry a reasonable approximation of what is happening and that these data should be developed to reflect events.

A good artificial intelligence engineer is confident across a broad range of expertise and against a rapidly changing environment in which the tools and methods used to pursue our profession are in continual flux. Most of what we do is safely hidden from view.

The one area where what we do rises to the awareness of the lay public is in the presentation of our results. It is also an area with continual development of new visualization tools and techniques.

This is to highlight that the presentation part of this course may date the fastest and you should take from it principles and approaches to presentation and not necessarily the software tools.

Presentation is everything from writing up academic findings for publication in a journal, to writing a financial and market report for a business, to producing journalism on a complex and fast-moving topic, and to persuading donors and humanitarian agencies to take a particular health or environmental threat seriously.

It is, first and foremost, about organizing your thoughts to tell a consistent and compelling story.

8.3   A Language and Approach to Data-Driven Story-Telling

There are "lies, damned lies, and statistics," as Mark Twain used to say. Be very careful that you tell the story that is there, rather than one which reflects your own biases.

According to Edward Tufte, professor of statistics at Yale, graphical displays should:

• Show the data

- Induce the viewer to think about the substance, rather than about the methodology, graphic design, the technology of graphic production, or something else
- Avoid distorting what the data have to say
- Present many numbers in a small space
- Make large datasets coherent
- Encourage the eye to compare different pieces of data
- Reveal the data at several levels of detail, from a broad overview to the fine structure
- Serve a reasonably clear purpose: description, exploration, tabulation, or decoration
- Be closely integrated with the statistical and verbal descriptions of a dataset

There are a lot of people with a great many opinions about what constitutes good visual practice. Manual Lima, in his Visual Complexity blog, has even come up with an Information Visualization Manifesto.

Any story has a beginning, a middle, and a conclusion. The story-telling form can vary, but the best and most memorable stories have compelling narratives easily retold.

Throwing data at a bunch of charts in the hopes that something will stick does not promote engagement anymore than randomly plunking at an instrument produces music.

Story-telling does not just happen.

Sun Tzu said, "There are not more than five musical notes, yet the combinations of these five give rise to more melodies than can ever be heard."

These are the fundamental chart-types which are used in the course of our careers:

- Line chart
- Bar chart
- Stacked / area variations of bar and line
- Bubble-charts
- Text charts
- Choropleth maps
- Tree maps

In addition, we can use small multiple versions of any of the above to enhance comparisons. Small multiples are simple charts placed alongside each other in a way that encourages analysis while still telling an engaging story. The axes are the same throughout and extraneous chart guides (like dividers between the charts and the vertical axes) have been removed. The simple line chart becomes both modern and information-dense when presented in this way.

There are numerous special types of charts (such as Chernoff Faces), but you are unlikely to have these implemented in your charting software.

Here is a simple methodology for developing a visual story:

- Write a flow-chart of the narrative encapsulating each of the components in a module.

- Each module will encapsulate a single data-driven *thought* and the type of chart will be imposed by the data:
  - Time series can be presented in line charts or by small multiples of other plots
  - Geospatial data invites choropleths
  - Complex multivariate data can be presented in tree maps
- In all matters, be led by the data and by good sense.
- Arrange those modules in a series of illustrations.
- Revise and edit according to the rules in the previous points.

Writing a narrative dashboard with multiple charts can be guided by George Orwell's rules from *Politics and the English Language*:

1. Never use a pie chart; use a table instead.
2. Never use a complicated chart where a simple one will do.
3. Never clutter your data with unnecessary grids, ticks, labels, or detail.
4. If it is possible to remove a chart without taking away from your story, always remove it.
5. Never mislead your reader through confusing or ambiguous axes or visualizations.
6. Break any of these rules sooner than draw anything outright barbarous.

8.4   Example: Telling Story with Data

Our example would be to telling the story of an epidemic in Yemen.

We have covered a great deal in this first lesson and now we come to the final section. Before we go further, we need two new libraries. *GeoPandas* is almost identical to Pandas but permits us to work with geospatial data (of which, more in a moment). *Seaborn* is similar to Matplotlib (and is a simplified wrapper around Matplotlib) but looks better, is designed for statistical data, and is simpler to use.

Our first step is to improve the line chart drawn at the end of the *initial exploration*. I mentioned the notion of *small multiples* earlier, and here is our first opportunity to draw it. Notice how much can be achieved in only a few lines of code, most of which (below) is about formatting the charts themselves.

```
# Seaborn for plotting and styling
import seaborn as sns

# Everything you need to know about Seaborn FacetGrid
# https://seaborn.pydata.org/generated/seaborn.FacetGrid.html
↪#seaborn.FacetGrid
sm = sns.FacetGrid(data_slice, col="Governorate", col_wrap=5,␣
↪size=3, aspect=2, margin_titles=True)
sm = sm.map(plt.plot, "Date", "Cases")
```

(continues on next page)

(continued from previous page)

```
# And now format the plots with appropriate titles and font
↪sizes
sm.set_titles("{col_name}", size=22).set_ylabels(size=20).set_
↪yticklabels(size=15).set_xlabels(size=20).set_
↪xticklabels(size=12)
```

```
C:\Users\turuk\Anaconda3\envs\calabar\lib\site-packages\
↪seaborn\axisgrid.py:230: UserWarning: The `size` parameter
↪has been renamed to `height`; please update your code.
  warnings.warn(msg, UserWarning)
```

```
<seaborn.axisgrid.FacetGrid at 0x1a22f3e5b70>
```

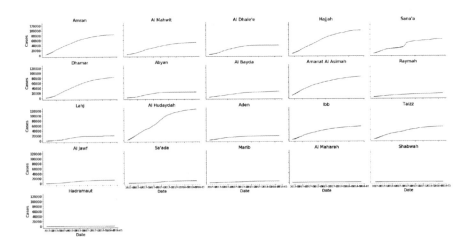

Notice how, even with the condensed format, it is still straightforward to understand what is happening and the overall display makes for a compelling and engaging visual.

Unfortunately, unless you know Yemen well, this story is incomplete. It is difficult to see where these changes are taking place, or how each governorate is related to the others in physical space. For that we need to plot our data onto a map.

There are a number of limits for publishing data on maps:

- A choropleth map is really a type of bar chart where the height of the bars is reflected by a color gradient in 2D space.
- Boundaries that make up regions, districts (or governorates) are of wildly different sizes and can mislead into prioritizing size over color scale.

Despite these limitations, map-based charts are useful for grounding data in a physical place. When used in combination with other charts (such as the line charts above), one can build a complete narrative.

To draw a map, we need a shapefile. These are a collection of several files developed according to a standard created by Esri that contain information shapes defined by geographic points, polylines, or polygons, as well as additional files with metadata or attributes.

HDX has exactly what we need as *Yemen—Administrative Boundaries.* Download the shapefile zip files to a folder and unzip all the files.

Now we are going to create a GeoPandas dataframe to open the shapefile and then join this dataframe to our existing data so that we can draw maps.

```
# Import our GeoPandas library
import geopandas as gpd
# Open the shapefile called "yem_admin1.shp" and note that -
↪if you're doing this on your home
# computer, you'll need to load the file from where-ever you
↪saved it
shape_data = gpd.GeoDataFrame.from_file("data/yem_admin1.shp")
# We have no data for Socotra island, so we can drop this row
shape_data = shape_data.loc[~shape_data.name_en.isin(["Socotra
↪"])]
# And now we can merge our existing data_slice to produce our
↪map data
map_data = pd.merge(shape_data, data_slice, how="outer", left_
↪on="name_en", right_on="Governorate", indicator=False)

# Let's draw a map

# First, define a figure, axis and plot size
fig, ax = plt.subplots(figsize=(25,14))
# We'll look at one specific date, the last entry in the series
md = map_data.loc[map_data.Date == "2018-01-14"]
# And plot
md.plot(ax=ax, column='Cases', cmap='OrRd')
```

```
<matplotlib.axes._subplots.AxesSubplot at 0x1cb20727cc0>
```

And here we hit a fundamental limit of a map; it would be nice to show a time series of how events progressed.

Well, remember the small multiple. So, to end this first lesson, here is what a small multiple map looks like.

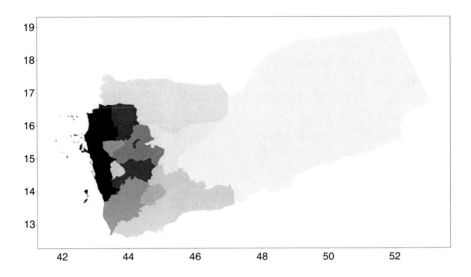

```
# This is a bit more complex than you may expect ... but think␣
↪of it like this:
# We're going to create a figure and then iterate over the␣
↪time-series to progressively
# add in new subplots. Since there are 125 dates - and that's␣
↪rather a lot - we'll
# deliberately limit this to the first date in each month, and␣
↪the final date.

# Create a datetime format data series
dates = pd.Series([pd.Timestamp(d) for d in map_data["Date"].
↪unique()])
# Sort dates in place
dates.sort_values(inplace = True)
dl = {}
for d in dates:
    # A mechanism to get the last day of each year-month
    k = "{}-{}".format(d.year, d.month)
    dl[k] = d
# Recover and sort the unique list of dates
dates = list(dl.values())
dates.sort()

# Create our figure
fig = plt.figure(figsize=(18,10))
# Set two check_sums, first_date and sub_count
first_date = 0
sub_count = 1
# Loop through the dates, using "enumerate" to count the␣
↪number of times we loop
for i, d in enumerate(dates[:]):
```

(continues on next page)

(continued from previous page)

```
    # Get a dataframe for the subplot at this date
    subplot = map_data.loc[map_data.Date == d]
    # Add the appropriate subplot in a frame structured as 3␣
↪items in 3 rows
    # If you get errors here, it's probably because Matplotlib␣
↪was expecting
    # a different number of images that you are creating.␣
↪Check and adjust.
    ax = fig.add_subplot(3, 3, sub_count)
    # Increment the count
    sub_count+=1
    # Do some visual fixes to ensure we don't distort the maps,
↪ and provide titles
    ax.set_aspect('equal')
    ax.set_axis_off()
    ax.title.set_text(d.date())
    # And plot
    subplot.plot(ax=ax, column='Cases', cmap='OrRd')
```

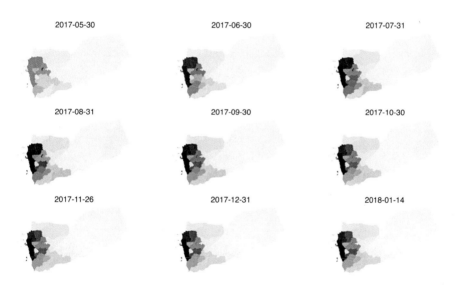

That brings us to the end of this lesson and this case-study. You can play around with the code, pick a different column to visualize (perhaps "Deaths"), and can learn more in the libraries about how to present these charts.

Chapter 9
Data Wrangling

Abstract Often, data collected from the source are messy and incomplete, which cannot be fed directly into Artificial Intelligence Programs. Data Wrangling skills are needed to create efficient ETL pipelines for usable data. There are many functions in Pandas that allow us to deal with a wide variety of circumstances. This chapter will illustrate how to handle Missing Data Values, Duplicates, Mapping Values, Outliers, Permutations, Merging and Combining, Reshaping, and Pivoting.

Learning outcomes:

- Learn how to use pandas to perform data cleaning and data wrangling.
- Apply data cleaning and data wrangling techniques on real life examples.

Suppose you are working on a machine learning project. You decide to use your favorite classification algorithm only to realize that the training dataset contains a mixture of continuous and categorical variables and you'll need to transform some of the variables into a suitable format. You realize that the raw data you have can't be used for your analysis without some manipulation—what you'll soon know as data wrangling. You'll need to clean this messy data to get anywhere with it.

It is often the case with data science projects that you'll have to deal with messy or incomplete data. The raw data we obtain from different data sources is often unusable at the beginning. All the activity that you do on the raw data to make it "clean" enough to input to your analytical algorithm is called data wrangling or data munging. If you want to create an efficient ETL pipeline (extract, transform, and load) or create beautiful data visualizations, you should be prepared to do a lot of data wrangling.

As most statisticians, data analysts, and data scientists will admit, most of the time spent implementing an analysis is devoted to cleaning or wrangling the data itself, rather than to coding or running a particular model that uses the data. According to O'Reilly's 2016 Data Science Salary Survey, 69% of data scientists will spend a significant amount of time in their day-to-day dealing with basic exploratory data analysis, while 53% spend time cleaning their data. Data wrangling is an essential part of the data science role—and if you gain data wrangling skills and become proficient at it, you'll quickly be recognized as somebody who can

© The Author(s), under exclusive license to Springer Nature Singapore Pte Ltd. 2022   149
T. T. Teoh, Z. Rong, *Artificial Intelligence with Python,*
Machine Learning: Foundations, Methodologies, and Applications,
https://doi.org/10.1007/978-981-16-8615-3_9

contribute to cutting-edge data science work and who can hold their own as a data professional.

In this chapter, we will be implementing and showing some of the most common data wrangling techniques used in the industry. But first, let us import the required libraries.

```
import numpy as np
import pandas as pd
import matplotlib.pyplot as plt
```

9.1   Handling Missing Data

9.1.1   Missing Data

```
string_data = pd.Series(['aardvark', 'artichoke', np.nan,
↪'avocado'])
string_data
string_data.isnull()
```

```
0      False
1      False
2       True
3      False
dtype: bool
```

```
string_data[0] = None
string_data.isnull()
```

```
0       True
1      False
2       True
3      False
dtype: bool
```

9.1.2   Removing Missing Data

```
from numpy import nan as NA
data = pd.Series([1, NA, 3.5, NA, 7])
data.dropna()
```

```
0    1.0
2    3.5
4    7.0
dtype: float64
```

```
data[data.notnull()]
```

```
0    1.0
2    3.5
4    7.0
dtype: float64
```

```
data = pd.DataFrame([[1., 6.5, 3.], [1., NA, NA],
                     [NA, NA, NA], [NA, 6.5, 3.]])
cleaned = data.dropna()
print(data)
print(cleaned)
```

```
     0    1    2
0  1.0  6.5  3.0
1  1.0  NaN  NaN
2  NaN  NaN  NaN
3  NaN  6.5  3.0
     0    1    2
0  1.0  6.5  3.0
```

```
data.dropna(how='all')
```

```
     0    1    2
0  1.0  6.5  3.0
1  1.0  NaN  NaN
3  NaN  6.5  3.0
```

```
data[4] = NA
data
data.dropna(axis=1, how='all')
```

```
     0    1    2
0  1.0  6.5  3.0
1  1.0  NaN  NaN
2  NaN  NaN  NaN
3  NaN  6.5  3.0
```

```
df = pd.DataFrame(np.random.randn(7, 3))
df.iloc[:4, 1] = NA
df.iloc[:2, 2] = NA
df
```

(continues on next page)

(continued from previous page)

```
df.dropna()
df.dropna(thresh=2)
```

```
          0          1          2
2  -0.185468        NaN  -1.250882
3  -0.250543        NaN  -0.038900
4  -1.658802  -1.346946   0.962846
5   0.439124  -1.433696  -0.169313
6   1.531410  -0.172615   0.203521
```

9.2   Transformation

9.2.1   *Duplicates*

```
data = pd.DataFrame({'k1': ['one', 'two'] * 3 + ['two'],
                     'k2': [1, 1, 2, 3, 3, 4, 4]})
data
```

```
    k1  k2
0  one   1
1  two   1
2  one   2
3  two   3
4  one   3
5  two   4
6  two   4
```

```
data.duplicated()
```

```
0    False
1    False
2    False
3    False
4    False
5    False
6     True
dtype: bool
```

```
data.drop_duplicates()
```

```
    k1  k2
0  one   1
1  two   1
```

(continues on next page)

(continued from previous page)

```
2   one     2
3   two     3
4   one     3
5   two     4
```

```
data['v1'] = range(7)
data.drop_duplicates(['k1'])
```

```
    k1  k2  v1
0   one  1   0
1   two  1   1
```

```
data.drop_duplicates(['k1', 'k2'], keep='last')
```

```
    k1  k2  v1
0   one  1   0
1   two  1   1
2   one  2   2
3   two  3   3
4   one  3   4
6   two  4   6
```

9.2.2 Mapping

```
data = pd.DataFrame({'food': ['bacon', 'pulled pork', 'bacon',
                              'Pastrami', 'corned beef', 'Bacon
↪',
                              'pastrami', 'honey ham', 'nova␣
↪lox'],
                     'ounces': [4, 3, 12, 6, 7.5, 8, 3, 5, 6]})
data
```

```
          food  ounces
0        bacon    4.0
1  pulled pork    3.0
2        bacon   12.0
3     Pastrami    6.0
4  corned beef    7.5
5        Bacon    8.0
6     pastrami    3.0
7    honey ham    5.0
8     nova lox    6.0
```

```
meat_to_animal = {
  'bacon': 'pig',
  'pulled pork': 'pig',
  'pastrami': 'cow',
  'corned beef': 'cow',
  'honey ham': 'pig',
  'nova lox': 'salmon'
}
```

```
lowercased = data['food'].str.lower()
lowercased
data['animal'] = lowercased.map(meat_to_animal)
data
```

```
          food   ounces   animal
0        bacon      4.0      pig
1  pulled pork      3.0      pig
2        bacon     12.0      pig
3     Pastrami      6.0      cow
4  corned beef      7.5      cow
5        Bacon      8.0      pig
6     pastrami      3.0      cow
7    honey ham      5.0      pig
8     nova lox      6.0   salmon
```

```
data['food'].map(lambda x: meat_to_animal[x.lower()])
```

```
0        pig
1        pig
2        pig
3        cow
4        cow
5        pig
6        cow
7        pig
8     salmon
Name: food, dtype: object
```

9.3   Outliers

```
data = pd.DataFrame(np.random.randn(1000, 4))
data.describe()
```

	0	1	2	3
count	1000.000000	1000.000000	1000.000000	1000.000000

(continues on next page)

(continued from previous page)

mean	0.046326	0.025065	-0.020071	0.028721
std	0.998348	0.985165	0.995451	1.032522
min	-3.422314	-3.266015	-2.954779	-3.222582
25%	-0.589264	-0.659314	-0.667673	-0.652942
50%	0.022320	0.034156	-0.019490	0.035827
75%	0.705115	0.700335	0.615950	0.709712
max	3.455663	3.191903	2.767412	3.355966

```
col = data[2]
col[np.abs(col) > 3]
```

```
Series([], Name: 2, dtype: float64)
```

```
data[(np.abs(data) > 3).any(1)]
```

	0	1	2	3
193	-0.051466	-1.147485	0.704028	-3.222582
263	2.092650	-3.266015	0.249550	1.422404
509	-0.552704	-1.032550	-0.980024	3.355966
592	1.297188	3.191903	-0.459355	1.490715
612	-3.422314	-1.407894	-0.076225	-2.017783
640	-3.254393	-0.378483	-1.233516	0.040324
771	3.167948	-0.128717	-0.809991	-1.400584
946	3.455663	-1.112744	-1.017207	1.736736
973	2.014649	0.441878	-1.071450	-3.103078
983	-1.566632	-3.011891	0.161519	-0.468655

```
data[np.abs(data) > 3] = np.sign(data) * 3
data.describe()
```

	0	1	2	3
count	1000.000000	1000.000000	1000.000000	1000.000000
mean	0.046380	0.025151	-0.020071	0.028690
std	0.994186	0.983675	0.995451	1.030448
min	-3.000000	-3.000000	-2.954779	-3.000000
25%	-0.589264	-0.659314	-0.667673	-0.652942
50%	0.022320	0.034156	-0.019490	0.035827
75%	0.705115	0.700335	0.615950	0.709712
max	3.000000	3.000000	2.767412	3.000000

```
np.sign(data).head()
```

	0	1	2	3
0	1.0	-1.0	1.0	1.0
1	-1.0	1.0	-1.0	1.0
2	1.0	-1.0	1.0	1.0

(continues on next page)

(continued from previous page)

```
3 -1.0 -1.0  1.0 -1.0
4  1.0 -1.0 -1.0 -1.0
```

9.4   Permutation

```
df = pd.DataFrame(np.arange(5 * 4).reshape((5, 4)))
sampler = np.random.permutation(5)
sampler
```

```
array([2, 0, 3, 1, 4])
```

```
df
df.take(sampler)
```

```
     0   1   2   3
2    8   9  10  11
0    0   1   2   3
3   12  13  14  15
1    4   5   6   7
4   16  17  18  19
```

```
df.sample(n=3)
```

```
     0   1   2   3
3   12  13  14  15
1    4   5   6   7
0    0   1   2   3
```

```
choices = pd.Series([5, 7, -1, 6, 4])
draws = choices.sample(n=10, replace=True)
draws
```

```
2    -1
0     5
3     6
3     6
2    -1
0     5
0     5
2    -1
0     5
1     7
dtype: int64
```

9.5 Merging and Combining

```
df1 = pd.DataFrame({'key': ['b', 'b', 'a', 'c', 'a', 'a', 'b'],
                    'data1': range(7)})
df2 = pd.DataFrame({'key': ['a', 'b', 'd'],
                    'data2': range(3)})
df1
df2
```

```
   key  data2
0   a      0
1   b      1
2   d      2
```

```
pd.merge(df1, df2)
```

```
   key  data1  data2
0   b      0      1
1   b      1      1
2   b      6      1
3   a      2      0
4   a      4      0
5   a      5      0
```

```
pd.merge(df1, df2, on='key')
```

```
   key  data1  data2
0   b      0      1
1   b      1      1
2   b      6      1
3   a      2      0
4   a      4      0
5   a      5      0
```

```
df3 = pd.DataFrame({'lkey': ['b', 'b', 'a', 'c', 'a', 'a', 'b
↪'],
                    'data1': range(7)})
df4 = pd.DataFrame({'rkey': ['a', 'b', 'd'],
                    'data2': range(3)})
pd.merge(df3, df4, left_on='lkey', right_on='rkey')
```

```
   lkey  data1 rkey  data2
0    b      0    b      1
1    b      1    b      1
2    b      6    b      1
3    a      2    a      0
4    a      4    a      0
5    a      5    a      0
```

```
pd.merge(df1, df2, how='outer')
```

```
   key  data1  data2
0    b    0.0    1.0
1    b    1.0    1.0
2    b    6.0    1.0
3    a    2.0    0.0
4    a    4.0    0.0
5    a    5.0    0.0
6    c    3.0    NaN
7    d    NaN    2.0
```

```
df1 = pd.DataFrame({'key': ['b', 'b', 'a', 'c', 'a', 'b'],
                    'data1': range(6)})
df2 = pd.DataFrame({'key': ['a', 'b', 'a', 'b', 'd'],
                    'data2': range(5)})
df1
df2
pd.merge(df1, df2, on='key', how='left')
```

```
    key  data1  data2
0     b      0    1.0
1     b      0    3.0
2     b      1    1.0
3     b      1    3.0
4     a      2    0.0
5     a      2    2.0
6     c      3    NaN
7     a      4    0.0
8     a      4    2.0
9     b      5    1.0
10    b      5    3.0
```

```
pd.merge(df1, df2, how='inner')
```

```
   key  data1  data2
0    b      0      1
1    b      0      3
2    b      1      1
3    b      1      3
4    b      5      1
5    b      5      3
6    a      2      0
7    a      2      2
8    a      4      0
9    a      4      2
```

```
left = pd.DataFrame({'key1': ['foo', 'foo', 'bar'],
                     'key2': ['one', 'two', 'one'],
```

(continues on next page)

(continued from previous page)

```
                        'lval': [1, 2, 3]})
right = pd.DataFrame({'key1': ['foo', 'foo', 'bar', 'bar'],
                        'key2': ['one', 'one', 'one', 'two'],
                        'rval': [4, 5, 6, 7]})
pd.merge(left, right, on=['key1', 'key2'], how='outer')
```

```
  key1 key2  lval  rval
0  foo  one   1.0   4.0
1  foo  one   1.0   5.0
2  foo  two   2.0   NaN
3  bar  one   3.0   6.0
4  bar  two   NaN   7.0
```

```
pd.merge(left, right, on='key1')
pd.merge(left, right, on='key1', suffixes=('_left', '_right'))
```

```
  key1 key2_left  lval key2_right  rval
0  foo       one     1        one     4
1  foo       one     1        one     5
2  foo       two     2        one     4
3  foo       two     2        one     5
4  bar       one     3        one     6
5  bar       one     3        two     7
```

9.6 Reshaping and Pivoting

```
data = pd.DataFrame(np.arange(6).reshape((2, 3)),
                    index=pd.Index(['Ohio', 'Colorado'], name=
↪'state'),
                    columns=pd.Index(['one', 'two', 'three'],
                    name='number'))
data
```

```
number     one  two  three
state
Ohio         0    1      2
Colorado     3    4      5
```

```
result = data.stack()
result
```

```
state      number
Ohio       one         0
```

(continues on next page)

(continued from previous page)

```
            two            1
            three          2
Colorado    one            3
            two            4
            three          5
dtype: int32
```

```
result.unstack()
```

```
number     one   two   three
state
Ohio         0     1       2
Colorado     3     4       5
```

```
result.unstack(0)
result.unstack('state')
```

```
state     Ohio   Colorado
number
one          0          3
two          1          4
three        2          5
```

```
s1 = pd.Series([0, 1, 2, 3], index=['a', 'b', 'c', 'd'])
s2 = pd.Series([4, 5, 6], index=['c', 'd', 'e'])
data2 = pd.concat([s1, s2], keys=['one', 'two'])
data2
data2.unstack()
```

```
        a      b      c      d      e
one   0.0    1.0    2.0    3.0    NaN
two   NaN    NaN    4.0    5.0    6.0
```

9.7   Wide to Long

```
df = pd.DataFrame({'key': ['foo', 'bar', 'baz'],
                   'A': [1, 2, 3],
                   'B': [4, 5, 6],
                   'C': [7, 8, 9]})
df
```

```
    key   A   B   C
0   foo   1   4   7
```

(continues on next page)

(continued from previous page)

```
1   bar   2   5   8
2   baz   3   6   9
```

```
melted = pd.melt(df, ['key'])
melted
```

```
    key variable   value
0   foo         A       1
1   bar         A       2
2   baz         A       3
3   foo         B       4
4   bar         B       5
5   baz         B       6
6   foo         C       7
7   bar         C       8
8   baz         C       9
```

```
reshaped = melted.pivot('key', 'variable', 'value')
reshaped
```

```
variable  A   B   C
key
bar       2   5   8
baz       3   6   9
foo       1   4   7
```

```
reshaped.reset_index()
```

```
variable  key   A   B   C
0         bar   2   5   8
1         baz   3   6   9
2         foo   1   4   7
```

```
pd.melt(df, id_vars=['key'], value_vars=['A', 'B'])
```

```
    key variable   value
0   foo         A       1
1   bar         A       2
2   baz         A       3
3   foo         B       4
4   bar         B       5
5   baz         B       6
```

```
pd.melt(df, value_vars=['A', 'B', 'C'])
pd.melt(df, value_vars=['key', 'A', 'B'])
```

```
   variable value
0       key   foo
1       key   bar
2       key   baz
3         A     1
4         A     2
5         A     3
6         B     4
7         B     5
8         B     6
```

Chapter 10
Regression

Abstract Regression estimates the relationship between dependent variables and independent variables. Linear regression is an easily understood, popular basic technique, which uses historical data to produce an output variable. Decision Tree regression arrives at an estimate by applying conditional rules on the data, narrowing possible values until a single prediction is made. Random Forests are clusters of individual decision trees that produce a prediction by selecting a vote by majority voting. Neural Networks are a representation of the brain and learns from the data through adjusting weights to minimize the error of prediction. Proper Data Processing techniques can further improve a model's prediction such as ranking feature importance and outlier removal.

Learning outcomes:

- Learn and apply basic models for regression tasks using sklearn and keras.
- Learn data processing techniques to achieve better results.
- Learn how to use simple feature selection techniques to improve our model.
- Data cleaning to help improve our model's RMSE

Regression looks for relationships among variables. For example, you can observe several employees of some company and try to understand how their salaries depend on the features, such as experience, level of education, role, city they work in, and so on.

This is a regression problem where data related to each employee represent one observation. The presumption is that the experience, education, role, and city are the independent features, and the salary of the employee depends on them.

Similarly, you can try to establish a mathematical dependence of the prices of houses on their areas, numbers of bedrooms, distances to the city center, and so on.

Generally, in regression analysis, you usually consider some phenomenon of interest and have a number of observations. Each observation has two or more features. Following the assumption that (at least) one of the features depends on the others, you try to establish a relation among them.

The dependent features are called the dependent variables, outputs, or responses.

T. T. Teoh, Z. Rong, *Artificial Intelligence with Python*,
Machine Learning: Foundations, Methodologies, and Applications,
https://doi.org/10.1007/978-981-16-8615-3_10

The independent features are called the independent variables, inputs, or predictors.

Regression problems usually have one continuous and unbounded dependent variable. The inputs, however, can be continuous, discrete, or even categorical data such as gender, nationality, brand, and so on.

It is a common practice to denote the outputs with x and inputs with y. If there are two or more independent variables, they can be represented as the vector $x = (x_1, \ldots, x_r)$, where r is the number of inputs.

When Do You Need Regression?
Typically, you need regression to answer whether and how some phenomenon influences the other or how several variables are related. For example, you can use it to determine if and to what extent the experience or gender impacts salaries.

Regression is also useful when you want to forecast a response using a new set of predictors. For example, you could try to predict electricity consumption of a household for the next hour given the outdoor temperature, time of day, and number of residents in that household.

Regression is used in many different fields: economy, computer science, social sciences, and so on. Its importance rises every day with the availability of large amounts of data and increased awareness of the practical value of data.

It is important to note is that regression does not imply causation. It is easy to find examples of non-related data that, after a regression calculation, do pass all sorts of statistical tests. The following is a popular example that illustrates the concept of data-driven "causality."

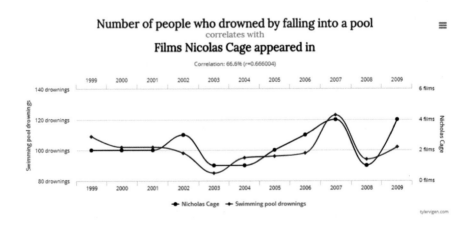

It is often said that correlation does not imply causation, although, inadvertently, we sometimes make the mistake of supposing that there is a causal link between two variables that follow a certain common pattern

Dataset: "Alumni Giving Regression (Edited).csv"
You can obtain the dataset from this link:

```
https://www.dropbox.com/s/veak3ugc4wj9luz/Alumni%20Giving
↪%20Regression%20%28Edited%29.csv?dl=0.
```

Also, you may run the following code in order to download the dataset in
`google colab`:

```
!wget https://www.dropbox.com/s/veak3ugc4wj9luz/Alumni%20Giving
↪%20Regression%20%28Edited%29.csv?dl=0 -O
--quiet  "./Alumni Giving Regression (Edited).csv"
```

```
!wget https://www.dropbox.com/s/veak3ugc4wj9luz/Alumni%20Giving
↪%20Regression%20%28Edited%29.csv?dl=0 -O -quiet  "./Alumni␣
↪Giving Regression (Edited).csv"
```

```
# Importing libraries needed
# Note that keras is generally used for deep learning as well
from keras.models import Sequential
from keras.layers import Dense, Dropout
from sklearn.metrics import classification_report, confusion_
↪matrix
from sklearn.model_selection import train_test_split
from sklearn.metrics import mean_squared_error
import numpy as np
from sklearn import linear_model
from sklearn import preprocessing
from sklearn import tree
from sklearn.ensemble import RandomForestRegressor,
↪GradientBoostingRegressor
import pandas as pd
import csv
```

```
Using TensorFlow backend.
```

In general, we will import dataset for structured dataset using pandas. We will
also demonstrate the code for loading dataset using NumPy to show the differences
between both libraries. Here, we are using a method in pandas call `read_csv`,
which takes the path of a csv file. `'CS'` in CSV represents comma separated. Thus,
if you open up the file in excel, you would see values separated by commas.

```
# fix random seed for reproducibility
np.random.seed(7)
df = pd.read_csv("Alumni Giving Regression (Edited).csv",␣
↪delimiter="," )
df.head()
```

	A	B	C	D	E	F
0	24	0.42	0.16	0.59	0.81	0.08
1	19	0.49	0.04	0.37	0.69	0.11
2	18	0.24	0.17	0.66	0.87	0.31
3	8	0.74	0.00	0.81	0.88	0.11
4	8	0.95	0.00	0.86	0.92	0.28

In pandas, it is very convenient to handle numerical data. Before doing any model, it is good to take a look at some of the dataset's statistics to get a "feel" of the data. Here, we can simple call df.describe, which is a method in pandas dataframe

```
df.describe()
```

	A	B	C	D	⌙
↪ E	F				
count	123.000000	123.000000	123.000000	123.000000	123.
↪000000	123.000000				
mean	17.772358	0.403659	0.136260	0.645203	0.
↪841138	0.141789				
std	4.517385	0.133897	0.060101	0.169794	0.
↪083942	0.080674				
min	6.000000	0.140000	0.000000	0.260000	0.
↪580000	0.020000				
25%	16.000000	0.320000	0.095000	0.505000	0.
↪780000	0.080000				
50%	18.000000	0.380000	0.130000	0.640000	0.
↪840000	0.130000				
75%	20.000000	0.460000	0.180000	0.785000	0.
↪910000	0.170000				
max	31.000000	0.950000	0.310000	0.960000	0.
↪980000	0.410000				

Furthermore, pandas provides a helpful method to calculate the pairwise correlation between two variables. What is correlation?

The term "correlation" refers to a mutual relationship or association between quantities (numerical number). In almost any business, it is very helping to express one quantity in terms of its relationship with others. We are concerned with this because business plans and departments are not isolated! For example, sales might increase when the marketing department spends more on advertisements, or a customer's average purchase amount on an online site may depend on his or her characteristics. Often, correlation is the first step to understanding these relationships and subsequently building better business and statistical models.

For example: "D" and "E" have a strong correlation of 0.93, which means that when D moves in the positive direction E is likely to move in that direction too. Here, notice that the correlation of A and A is 1. Of course, A would be perfectly correlated with A.

```
corr=df.corr(method ='pearson')
corr
```

```
          A          B          C          D          E          F
A   1.000000  -0.691900   0.414978  -0.604574  -0.521985  -0.549244
B  -0.691900   1.000000  -0.581516   0.487248   0.376735   0.540427
C   0.414978  -0.581516   1.000000   0.017023   0.055766  -0.175102
D  -0.604574   0.487248   0.017023   1.000000   0.934396   0.681660
E  -0.521985   0.376735   0.055766   0.934396   1.000000   0.647625
F  -0.549244   0.540427  -0.175102   0.681660   0.647625   1.000000
```

In general, we would need to test our model. `train_test_split` is a function in `Sklearn` model selection for splitting data arrays into two subsets for training data and for testing data. With this function, you do not need to divide the dataset manually. You can use from the function `train_test_split` using the following code `sklearn.model_selection import train_test_split`. By default, Sklearn train_test_split will make random partitions for the two subsets. However, you can also specify a random state for the operation.

Here, take note that we will need to pass in the X and Y to the function. X refers to the `features` while Y refers to the `target` of the dataset.

```
Y_POSITION = 5
model_1_features = [i for i in range(0,Y_POSITION)]
X = df.iloc[:,model_1_features]
Y = df.iloc[:,Y_POSITION]
# create model
X_train, X_test, y_train, y_test = train_test_split(X, Y, test_
↪size=0.20, random_state=2020)
```

10.1 Linear Regression

Linear regression is a basic predictive analytics technique that uses historical data to predict an output variable. It is popular for predictive modeling because it is easily understood and can be explained using plain English.

The basic idea is that if we can fit a linear regression model to observed data, we can then use the model to predict any future values. For example, let us assume that we have found from historical data that the price (P) of a house is linearly dependent upon its size (S)—in fact, we found that a house's price is exactly 90 times its size. The equation will look like this: `P = 90*S`

With this model, we can then predict the cost of any house. If we have a house that is 1,500 square feet, we can calculate its price to be: `P = 90*1500 = $135,000`

There are two kinds of variables in a linear regression model:

- The input or predictor variable is the variable(s) that help predict the value of the output variable. It is commonly referred to as X.
- The output variable is the variable that we want to predict. It is commonly referred to as Y.

To estimate Y using linear regression, we assume the equation: $Y_e = \alpha + \beta\ X$ where Y_e is the estimated or predicted value of Y based on our linear equation. Our goal is to find statistically significant values of the parameters α and β that minimize the difference between Y and Y_e. If we are able to determine the optimum values of these two parameters, then we will have the line of best fit that we can use to predict the values of Y, given the value of X. So, how do we estimate α and β? We can use a method called ordinary least squares.

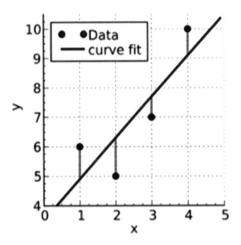

The objective of the least squares method is to find values of α and β that minimize the sum of the squared difference between Y and Y_e. We will not delve into the mathematics of least squares in our book.

Here, we notice that when E increases by 1, our Y increases by 0.175399. Also, when C increases by 1, our Y falls by 0.044160.

```
#Model 1 : linear regression

model1 = linear_model.LinearRegression()
model1.fit(X_train, y_train)
y_pred_train1 = model1.predict(X_train)
print("Regression")
print("================================")
RMSE_train1 = mean_squared_error(y_train,y_pred_train1)

print("Regression Train set: RMSE {}".format(RMSE_train1))
```

(continues on next page)

(continued from previous page)

```
print("================================")
y_pred1 = model1.predict(X_test)
RMSE_test1 = mean_squared_error(y_test,y_pred1)
print("Regression Test set: RMSE {}".format(RMSE_test1))
print("================================")

coef_dict = {}
for coef, feat in zip(model1.coef_,model_1_features):
    coef_dict[df.columns[feat]] = coef

print(coef_dict)
```

```
Regression
================================
Regression Train set: RMSE 0.0027616933222892287
================================
Regression Test set: RMSE 0.0042098240263563754
================================
{'A': -0.0009337757382417014, 'B': 0.16012156890162915, 'C': -
↪0.04416001542534971, 'D': 0.15217907817100398, 'E': 0.
↪17539950794101034}
```

10.2 Decision Tree Regression

A decision tree is arriving at an estimate by asking a series of questions to the data, each question narrowing our possible values until the model gets confident enough to make a single prediction. The order of the question and their content are being determined by the model. In addition, the questions asked are all in a True/False form.

This is a little tough to grasp because it is not how humans naturally think, and perhaps the best way to show this difference is to create a real decision tree from. In the above problem x1, x2 are two features that allow us to make predictions for the target variable y by asking True/False questions.

The decision of making strategic splits heavily affects a tree's accuracy. The decision criteria are different for classification and regression trees. Decision trees regression normally use mean squared error (MSE) to decide to split a node into two or more sub-nodes. Suppose we are doing a binary tree; the algorithm will first pick a value and split the data into two subsets. For each subset, it will calculate the MSE separately. The tree chooses the value with results in smallest MSE value.

Let us examine how is Splitting Decided for Decision Trees Regressor in more detail. The first step to create a tree is to create the first binary decision. How are you going to do it?

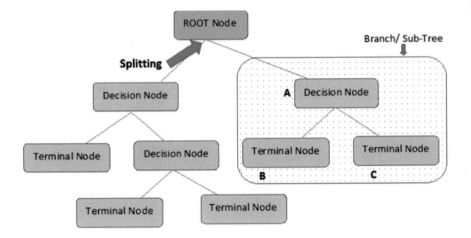

1. We need to pick a variable and the value to split on such that the two groups are as different from each other as possible.
2. For each variable, for each possible value of the possible value of that variable see whether it is better.
3. Take weighted average of two new nodes (mse*num_samples).

To sum up, we now have:

- A single number that represents how good a split is, which is the weighted average of the mean squared errors of the two groups that create.
- A way to find the best split, which is to try every variable and to try every possible value of that variable and see which variable and which value gives us a split with the best score.

Training of a decision tree regressor will stop when some stopping condition is met:

1. When you hit a limit that was requested (for example: max_depth).
2. When your leaf nodes only have one thing in them (no further split is possible, MSE for the train will be zero but will overfit for any other set—not a useful model).

```
#Model 2 decision tree
model2 = tree.DecisionTreeRegressor()
model2.fit(X_train, y_train)
print("Decision Tree")
print("=============================")
y_pred_train2 = model2.predict(X_train)
RMSE_train2 = mean_squared_error(y_train,y_pred_train2)
print("Decision Tree Train set: RMSE {}".format(RMSE_train2))
print("=============================")
```

(continues on next page)

(continued from previous page)

```
y_pred_test2 = model2.predict(X_test)
RMSE_test2 = mean_squared_error(y_test,y_pred_test2)
print("Decision Tree Test set: RMSE {}".format(RMSE_test2))
print("================================")
```

```
Decision Tree
================================
Decision Tree Train set: RMSE 1.4739259778473743e-36
================================
Decision Tree Test set: RMSE 0.008496
================================
```

10.3   Random Forests

What is a Random Forest? And how does it differ from a Decision Tree?

The fundamental concept behind random forest is a simple but powerful one—the wisdom of crowds. In data science speak, the reason that the random forest model works so well is: A large number of relatively uncorrelated models (trees) operating as a committee will outperform any of the individual constituent models.

The low correlation between models is the key. Just like how investments with low correlations (like stocks and bonds) come together to form a portfolio that is greater than the sum of its parts, uncorrelated models can produce ensemble predictions that are more accurate than any of the individual predictions. The reason for this wonderful effect is that the trees protect each other from their individual errors (as long as they do not constantly all err in the same direction). While some trees may be wrong, many other trees will be right, so as a group the trees are able to move in the correct direction. So the prerequisites for random forest to perform well are:

1. There needs to be some actual signals in our features so that models built using those features do better than random guessing.
2. The predictions (and therefore the errors) made by the individual trees need to have low correlations with each other.

So how does random forest ensure that the behavior of each individual tree is not too correlated with the behavior of any of the other trees in the model? It uses the following two methods:

1. Bagging (Bootstrap Aggregation)—Decision trees are very sensitive to the data they are trained on—small changes to the training set can result in significantly different tree structures. Random forest takes advantage of this by allowing each individual tree to randomly sample from the dataset with replacement, resulting in different trees. This process is known as bagging.

2. Feature Randomness—In a normal decision tree, when it is time to split a node, we consider every possible feature and pick the one that produces the most separation between the observations in the left node vs. those in the right node. In contrast, each tree in a random forest can pick only from a random subset of features. This forces even more variation among the trees in the model and ultimately results in lower correlation across trees and more diversification.

As Random Forest is actually a collection of Decision Trees, this makes the algorithm slower and less effective for real-time predictions. In general, Random Forest can be fast to train, but quite slow to create predictions once they are trained. This is due to the fact that it has to run predictions on each individual tree and then average their predictions to create the final prediction.

Each individual tree in the random forest splits out a class prediction and the class with the most votes becomes our model's prediction. Decision Trees do suffer from overfitting while Random Forest can prevent overfitting resulting in better prediction most of the time.

```
#Model 3 Random Forest
model3 = RandomForestRegressor()
model3.fit(X_train, y_train)
print("Random Forest Regressor")
print("===============================")
y_pred_train3 = model3.predict(X_train)
RMSE_train3 = mean_squared_error(y_train,y_pred_train3)
print("Random Forest Regressor TrainSet: RMSE {}".format(RMSE_
 ↪train3))
print("===============================")
y_pred_test3 = model3.predict(X_test)
RMSE_test3 = mean_squared_error(y_test,y_pred_test3)
print("Random Forest Regressor TestSet: RMSE {}".format(RMSE_
 ↪test3))
print("===============================")
```

```
Random Forest Regressor
===============================
Random Forest Regressor TrainSet: RMSE 0.0004964972448979589
===============================
Random Forest Regressor TestSet: RMSE 0.004843255999999997
===============================
```

10.4 Neural Network

Neural networks are the representation we make of the brain: neurons interconnected to other neurons, which forms a network. A simple information transits in a lot of them before becoming an actual thing, like "move the hand to pick up this pencil."

The operation of a complete neural network is straightforward : one enters variables as inputs (for example, an image if the neural network is supposed to tell what is on an image), and after some calculations, an output is returned (probability of whether an image is a cat).

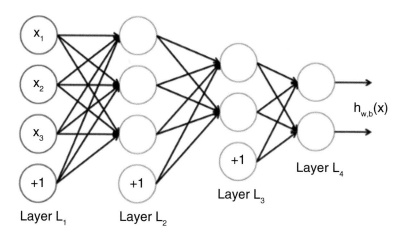

When an input is given to the neural network, it returns an output. On the first try, it cannot get the right output by its own (except with luck) and that is why, during the learning phase, every input comes with its label, explaining what output the neural network should have guessed. If the choice is the good one, actual parameters are kept and the next input is given. However, if the obtained output does not match the label, weights are changed. Those are the only variables that can be changed during the learning phase. This process may be imagined as multiple buttons that are turned into different possibilities every time an input is not guessed correctly. To determine which weight is better to modify, a particular process, called "backpropagation" is done.

Below is the code to create a simple neural network in python:

The following code is telling python to add a layer of 64 neurons into the neural network. We can stack the models by adding more layers of neuron. Or we can simply increase the number of neurons. This can be thought of as to increase the number of "neurons" in one's brain and thereby improving one's learning ability.

```
#Model 5: neural network
print("Neural Network")
print("===============================")
model = Sequential()
model.add(Dense(64, input_dim=Y_POSITION, activation='relu'))
model.add(Dense(64, activation='relu'))
model.add(Dropout(0.2))
model.add(Dense(1, activation='relu'))
# Compile mode
```

(continues on next page)

(continued from previous page)

```
# https://www.tensorflow.org/guide/keras/train_and_evaluate
model.compile(loss='MSE', optimizer='Adamax', metrics=[
 ↪'accuracy'])
# Fit the model
model.fit(X_train, y_train, epochs=300, batch_size=5,␣
 ↪verbose=0)
# evaluate the model
predictions5 = model.predict(X_train)
RMSE_train5 = mean_squared_error(y_train,predictions5)
print("Neural Network TrainSet: RMSE {}".format(RMSE_train5))
print("=================================")
predictions5 = model.predict(X_test)
RMSE_test5 = mean_squared_error(y_test,predictions5)
print("Neural Network TestSet: RMSE {}".format(RMSE_test5))
print("=================================")
```

```
Neural Network
=================================
Neural Network TrainSet: RMSE 0.02496122448979592
=================================
Neural Network TestSet: RMSE 0.032824
=================================
```

10.5   How to Improve Our Regression Model

10.5.1   Boxplot

A boxplot is a standardized way of displaying the distribution of data based on a
five number summary ("minimum," first quartile (Q1), median, third quartile (Q3),

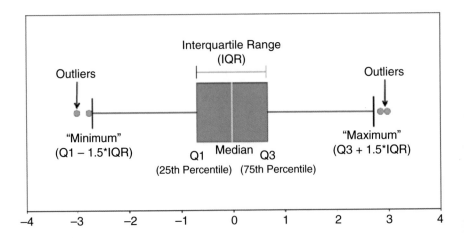

and "maximum"). It tells you about your outliers and what their values are. It can also tell you if your data is symmetrical, how tightly your data is grouped, and if and how your data is skewed.

Here is an image that shows normal distribution on a boxplot:

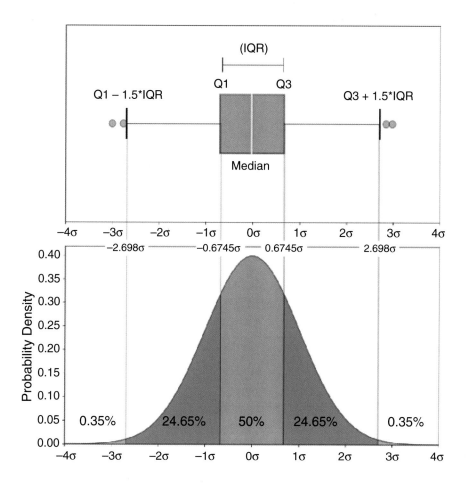

As seen, a boxplot is a great way to visualize your dataset. Now, let us try to remove the outliers using our boxplot plot. This can be easily achieved with pandas dataframe. But do note that the dataset should be numerical to do this.

Code for boxplot:

```
boxplot = pd.DataFrame(dataset).boxplot()
```

```
import seaborn as sns
import pandas as pd
boxplot = pd.DataFrame(dataset).boxplot()
```

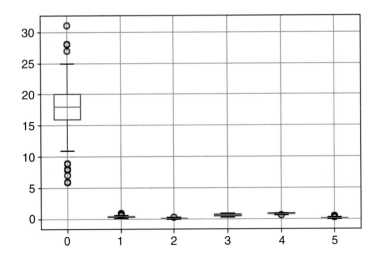

As shown in the plot, there are values in column 0 that are outliers, which are values that are extremely large or small. This can skew our dataset. A consequence of having outliers in our dataset is that our model cannot learn the right parameters. Thus, it results in a poorer prediction.

10.5.2 Remove Outlier

The code removes outlier that is more than 99th percentile. Next, let us apply this on values lower than 1st percentile.

```
quantile99 = df.iloc[:,0].quantile(0.99)
df1 = df[df.iloc[:,0] < quantile99]
df1.boxplot()
```

```
<matplotlib.axes._subplots.AxesSubplot at 0x1af88e4f108>
```

Here, we have removed the outliers from the data successfully.

```
quantile1 = df.iloc[:,0].quantile(0.01)
quantile99 = df.iloc[:,0].quantile(0.99)
df2 = df[(df.iloc[:,0] > quantile1) & (df.iloc[:,0] <
↪quantile99)]
df2.boxplot()
```

```
<matplotlib.axes._subplots.AxesSubplot at 0x1af8c38d308>
```

```
df2.shape
```

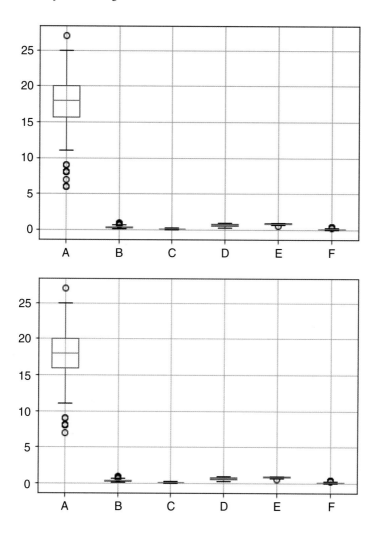

```
(118, 6)
```

10.5.3   Remove NA

To drop all the rows with the NaN values, you may use :

```
df.dropna()
```

```
df1 = df1.dropna()
```

10.6   Feature Importance

Apart from data cleaning, we can apply use variables that we deem to be important to us. One way of doing so is via feature importance of random forest trees. In many use cases it is equally important to not only have an accurate but also an interpretable model. Oftentimes, apart from wanting to know what our model's house price prediction is, we also wonder why it is this high/low and which features are most important in determining the forecast. Another example might be predicting customer churn—it is very nice to have a model that is successfully predicting which customers are prone to churn, but identifying which variables are important can help us in early detection and maybe even improving the product/service.

Knowing feature importance indicated by machine learning models can benefit you in multiple ways, for example:

1. By getting a better understanding of the model's logic you can not only verify it being correct but also work on improving the model by focusing only on the important variables.
2. The above can be used for variable selection—you can remove x variables that are not that significant and have similar or better performance in much shorter training time.
3. In some business cases it makes sense to sacrifice some accuracy for the sake of interpretability. For example, when a bank rejects a loan application, it must also have a reasoning behind the decision, which can also be presented to the customer.

We can obtain the feature importance using this code:

```
importances = RF.feature_importances_
```

Then, we can sort the feature importance for ranking and indexing.

```
indices = numpy.argsort(importances)[::-1]
```

```
import numpy
RF = model3
importances = RF.feature_importances_
std = numpy.std([tree.feature_importances_ for tree in RF.
    ↪estimators_],axis=0)
indices = numpy.argsort(importances)[::-1]

# Print the feature ranking
print("Feature ranking:")

for f in range(X.shape[1]):
    print("%d. feature (Column index) %s (%f)" % (f + 1,
    ↪indices[f], importances[indices[f]]))
```

```
Feature ranking:
1. feature (Column index) 3 (0.346682)
2. feature (Column index) 1 (0.217437)
3. feature (Column index) 0 (0.174081)
4. feature (Column index) 4 (0.172636)
5. feature (Column index) 2 (0.089163)
```

Let us use the top 3 features and retrain another model. Here, we took a shorter time to train the model, yet the RMSE does not suffer due to fewer features.

```
indices_top3 = indices[:3]
print(indices_top3)
dataset=df
df = pd.DataFrame(df)

Y_position = 5
TOP_N_FEATURE = 3

X = dataset.iloc[:,indices_top3]
Y = dataset.iloc[:,Y_position]
# create model
X_train, X_test, y_train, y_test = train_test_split(X, Y, test_
↪size=0.20, random_state=2020)
#Model 1 : linear regression


model1 = linear_model.LinearRegression()
model1.fit(X_train, y_train)
y_pred_train1 = model1.predict(X_train)
print("Regression")
print("===============================")
RMSE_train1 = mean_squared_error(y_train,y_pred_train1)

print("Regression TrainSet: RMSE {}".format(RMSE_train1))
print("===============================")
y_pred1 = model1.predict(X_test)
RMSE_test1 = mean_squared_error(y_test,y_pred1)
print("Regression Testset: RMSE {}".format(RMSE_test1))
print("===============================")
```

```
[3 1 0]
Regression
===============================
Regression TrainSet: RMSE 0.0027952079052752685
===============================
Regression Testset: RMSE 0.004341758028139643
===============================
```

10.7   Sample Code

```
import pandas as pd

df=pd.read_csv("C:/Users/User/Dropbox/TT Library/AI Model/
↪Python/Treynor (Regression).csv")
print(df)
```

```
df=df.dropna()
print(df)
```

```
# Split X, Y

X=df.iloc[:,0:len(df.columns)-1]
Y=df.iloc[:,len(df.columns)-1]

print(X)
print(Y)
```

```
#split train test

from sklearn.model_selection import train_test_split

X_train, X_test, Y_train, Y_test = train_test_split(X, Y, test_
↪size=0.3)
```

```
print(X_train)
print(X_test)
print(Y_train)
print(Y_test)
```

```
from sklearn import linear_model
from sklearn.metrics import mean_squared_error

model=linear_model.LinearRegression()
model.fit(X_train, Y_train)
pred=model.predict(X_train)
print(mean_squared_error(pred, Y_train))

pred=model.predict(X_test)
print(mean_squared_error(pred, Y_test))
```

```
model= linear_model.Ridge()
model.fit(X_train, Y_train)
pred=model.predict(X_train)
print(mean_squared_error(pred, Y_train))
```

(continues on next page)

(continued from previous page)

```
pred=model.predict(X_test)
print(mean_squared_error(pred, Y_test))
```

```
model= linear_model.Lasso()
model.fit(X_train, Y_train)
pred=model.predict(X_train)
print(mean_squared_error(pred, Y_train))

pred=model.predict(X_test)
print(mean_squared_error(pred, Y_test))
```

```
from sklearn import tree

model=tree.DecisionTreeRegressor()
model.fit(X_train, Y_train)
pred=model.predict(X_train)
print(mean_squared_error(pred, Y_train))

pred=model.predict(X_test)
print(mean_squared_error(pred, Y_test))
```

Chapter 11
Classification

Abstract Generally classification is about predicting a label whereas regression is usually used to predict a quantity. Classification models approximate a mapping function from inputs to produce a class or category. Logistic regression is used instead of linear regression for binary classification problems. It produces a value ranging from 0 to 1, which can be interpreted as a probability that an event occurred. Decision Trees, Random Forests, and Neural Networks can also be used for classification tasks, similar to that in regression. Support vector machines calculate a hyperplane that maximizes the distance between classes. Naive Bayes uses Bayes' Theorem to predict classes under the assumption of independence. Similarly, Data processing techniques can further improve classification performance as well.

Learning outcomes:

- Learn the difference between classification and regression. Be able to differentiate between classification and regression problems.
- Learn and apply basic models for classification tasks using sklearn and keras.
- Learn data processing techniques to achieve better classification results.

We have learnt about regression previously. Now, let us take a look at classification. Fundamentally, classification is about predicting a label and regression is about predicting a quantity.

Classification predictive modeling is the task of approximating a mapping function (f) from input variables (X) to discrete output variables (y). The output variables are often called labels or categories. The mapping function predicts the class or category for a given observation.

For example, an email of text can be classified as belonging to one of two classes: "spam" and "not spam." A classification can have real-valued or discrete input variables.

Here are different types of classification problem:

- A problem with two classes is often called a two-class or binary classification problem.

T. T. Teoh, Z. Rong, *Artificial Intelligence with Python*,
Machine Learning: Foundations, Methodologies, and Applications,
https://doi.org/10.1007/978-981-16-8615-3_11

- A problem with more than two classes is often called a multi-class classification problem.
- A problem where an example is assigned multiple classes is called a multi-label classification problem.

It is common for classification models to predict a continuous value as the probability of a given example belonging to each output class. The probabilities can be interpreted as the likelihood or confidence of a given example belonging to each class. A predicted probability can be converted into a class value by selecting the class label that has the highest probability.

For example, a specific email of text may be assigned the probabilities of 0.1 as being "spam" and 0.9 as being "not spam." We can convert these probabilities to a class label by selecting the "not spam" label as it has the highest predicted likelihood.

There are many ways to estimate the skill of a classification predictive model, but perhaps the most common is to calculate the classification accuracy.

The classification accuracy is the percentage of correctly classified examples out of all predictions made.

For example, if a classification predictive model made 5 predictions and 3 of them were correct and 2 of them were incorrect, then the classification accuracy of the model based on just these predictions would be

```
accuracy = correct predictions / total predictions * 100
accuracy = 3 / 5 * 100
accuracy = 60%
```

An algorithm that is capable of learning a classification predictive model is called a classification algorithm.

Dataset: "Diabetes (Edited).csv"

You can obtain the dataset from this link https://www.dropbox.com/s/ggxo241uog06yhj/Diabetes (Edited).csv?dl=0

Also, you may run the following code in order to download the dataset in google colab:

```
!wget https://www.dropbox.com/s/ggxo241uog06yhj/Diabetes%20
↪%28Edited%29.csv?dl=0 -O --quiet "Diabetes (Edited).csv"
```

```
from keras.models import Sequential
from keras.layers import Dense, Dropout
from sklearn.metrics import classification_report, confusion_
↪matrix
from sklearn.model_selection import train_test_split
import numpy
from sklearn import linear_model
from sklearn import preprocessing
```

(continues on next page)

(continued from previous page)

```
from sklearn import tree
from sklearn.ensemble import RandomForestClassifier,
↪GradientBoostingClassifier
import pandas as pd
import csv
```

Firstly, we will work on preprocessing the data. For numerical data, often we would preprocess the data by scaling it. In our example, we apply standard scalar, a popular preprocessing technique.

Standardization is a transformation that centers the data by removing the mean value of each feature and then scale it by dividing (non-constant) features by their standard deviation. After standardizing data the mean will be zero and the standard deviation one.

Standardization can drastically improve the performance of models. For instance, many elements used in the objective function of a learning algorithm assume that all features are centered around zero and have variance in the same order. If a feature has a variance that is orders of magnitude larger than others, it might dominate the objective function and make the estimator unable to learn from other features correctly as expected.

Here the code that does the scaling is as follows:

```
scaler = preprocessing.StandardScaler().fit(X_train)
scaled_X_train = scaler.transform(X_train)
scaled_X_test = scaler.transform(X_test)
```

Notice that we are using the scalar fitted on our X_train to transform values in X_test. This is to ensure that our model does not learn from the testing data. Usually, we would split our data before applying scaling. It is a bad practice to do scaling on the full dataset.

Apart from standard scaling we can use other scalar such as MinMaxScalar. feature_range refers to the highest and lowest values after scaling. By default, "feature_range" is −1 to 1. However, this range may prove to be too small as changes in our variable would be compressed to maximum of −1 to 1.

```
from sklearn.preprocessing import MinMaxScaler
scaler = MinMaxScaler(feature_range=(-3,3))
scaled_X_train = scaler.transform(X_train)
scaled_X_test = scaler.transform(X_test)
```

```
Y_position = 8

# fix random seed for reproducibility
numpy.random.seed(7)

df = pd.read_csv('Diabetes (Edited).csv')
print(df)
# summary statistics
```

(continues on next page)

(continued from previous page)

```
print(df.describe())

X = df.iloc[:,0:Y_position]
Y = df.iloc[:,Y_position]

# create model
X_train, X_test, y_train, y_test = train_test_split(X, Y, test_
↪size=0.40, random_state=2020)

#scaling to around -2 to 2 (Z)
scaler = preprocessing.StandardScaler().fit(X_train)
scaled_X_train = scaler.transform(X_train)
scaled_X_test = scaler.transform(X_test)
```

```
        A    B    C    D     E     F      G     H  I
0       6  148   72   35     0  33.6  0.627  50  1
1       1   85   66   29     0  26.6  0.351  31  0
2       8  183   64    0     0  23.3  0.672  32  1
3       1   89   66   23    94  28.1  0.167  21  0
4       0  137   40   35   168  43.1  2.288  33  1
..     ..  ...   ..   ..   ...   ...    ...  .. ..
763    10  101   76   48   180  32.9  0.171  63  0
764     2  122   70   27     0  36.8  0.340  27  0
765     5  121   72   23   112  26.2  0.245  30  0
766     1  126   60    0     0  30.1  0.349  47  1
767     1   93   70   31     0  30.4  0.315  23  0

[768 rows x 9 columns]
                A            B            C            D          ⌡
↪ E          F  \
count  768.000000   768.000000   768.000000   768.000000   768.
↪000000   768.000000
mean     3.845052   120.894531    69.105469    20.536458    79.
↪799479    31.992578
std      3.369578    31.972618    19.355807    15.952218   115.
↪244002     7.884160
min      0.000000     0.000000     0.000000     0.000000     0.
↪000000     0.000000
25%      1.000000    99.000000    62.000000     0.000000     0.
↪000000    27.300000
50%      3.000000   117.000000    72.000000    23.000000    30.
↪500000    32.000000
75%      6.000000   140.250000    80.000000    32.000000   127.
↪250000    36.600000
max     17.000000   199.000000   122.000000    99.000000   846.
↪000000    67.100000

                G            H            I
count  768.000000   768.000000   768.000000
mean     0.471876    33.240885     0.348958
std      0.331329    11.760232     0.476951
```

(continues on next page)

(continued from previous page)

min	0.078000	21.000000	0.000000
25%	0.243750	24.000000	0.000000
50%	0.372500	29.000000	0.000000
75%	0.626250	41.000000	1.000000
max	2.420000	81.000000	1.000000

In order to reduce code duplication as seen in the chapter on Regression. We can abstract the model and create a function to help us train and predict. Here is the explanation for the code:

```
model.fit(scaled_X_train, y_train)
```

We train the model using scaled_X_train and provide its label y_train

```
y_predicted = model3.predict(scaled_X_test)
```

We predict the model on our testing data and store its result in the variable y_predicted

```
cm_test = confusion_matrix(y_test,y_pred)
```

We create a confusion matrix given our y_test and y_pred. And what is a confusion matrix?

A Confusion matrix is an N x N matrix used for evaluating the performance of a classification model, where N is the number of target classes. The matrix compares the actual target values with those predicted by the model. This gives us a holistic view of how well our classification model is performing and what kinds of errors it is making.

- Expected down the side: Each row of the matrix corresponds to a predicted class.
- Predicted across the top: Each column of the matrix corresponds to an actual class.

```
acc_test = (cm_test[0,0] + cm_test[1,1]) / sum(sum(cm_test))
```

Lastly, this code calculates the accuracy for us. Accuracy is the number of correctly predicted data points out of all the data points. More formally, it is defined as the number of true positives and true negatives divided by the number of true positives, true negatives, false positives, and false negatives. These values are the outputs of a confusion matrix.

Here, we are assuming a binary classification problem. For multi-class classification problem, I would highly recommend using sklearn's accuracy function for its calculation.

```
def train_and_predict_using_model(model_name= "",model=None):
    model.fit(scaled_X_train, y_train)
    y_pred_train = model.predict(scaled_X_train)
    cm_train = confusion_matrix(y_train,y_pred_train)
```

(continues on next page)

(continued from previous page)

```
    print(model_name)
    print("=================================")
    print("Training confusion matrix: ")
    print(cm_train)
    acc_train = (cm_train[0,0] + cm_train[1,1]) / sum(sum(cm_
↪train))
    print("TrainSet: Accurarcy %.2f%%" % (acc_train*100))
    print("=================================")
    y_pred = model.predict(scaled_X_test)
    cm_test = confusion_matrix(y_test,y_pred)
    print(cm_test)
    acc_test = (cm_test[0,0] + cm_test[1,1]) / sum(sum(cm_
↪test))
    print("Testset: Accurarcy %.2f%%" % (acc_test*100))
    print("=================================")
```

11.1   Logistic Regression

Why not use linear regression?

Suppose we have a data of tumor size vs. its malignancy. As it is a classification problem, if we plot, we can see, all the values will lie on 0 and 1. And if we fit best found regression line, by assuming the threshold at 0.5, we can do line pretty reasonable job.

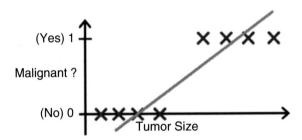

We can decide the point on the x axis from where all the values lying to its left side are considered as negative class and all the values lying to its right side are positive class.

But what if there is an outlier in the data. Things would get pretty messy. For example, for 0.5 threshold,

If we fit best found regression line, it still will not be enough to decide any point by which we can differentiate classes. It will put some positive class examples into negative class. The green dotted line (Decision Boundary) is dividing malignant tumors from benign tumors, but the line should have been at a yellow line that is clearly dividing the positive and negative examples. So just a single outlier

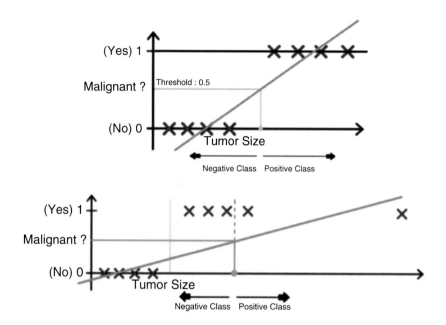

is disturbing the whole linear regression predictions. And that is where logistic regression comes into a picture.

As discussed earlier, to deal with outliers, Logistic Regression uses Sigmoid function. An explanation of logistic regression can begin with an explanation of the standard logistic function. The logistic function is a Sigmoid function, which takes any real value between zero and one. It is defined as

$$\sigma(t) = \frac{e^t}{e^t + 1} = \frac{1}{1 + e^{-t}}$$

And if we plot it, the graph will be S curve.

Now, when logistic regression model come across an outlier, it will take care of it.

Another way of looking at logistic regression:

Consider the case where we are looking at a classification problem and our output is probability. Our output is from 0 to 1, which represents the probability that the event has occurred. Using linear regression would result in output from 1 to infinity, which when mapped to a `sigmoid` function goes very well into 0 to 1 depending on the output of linear regression.

```
#https://scikit-learn.org/stable/modules/generated/sklearn.
↪linear_model.LogisticRegression.html
linear_classifier = linear_model.LogisticRegression(random_
↪state=123)
linear_classifier.fit(scaled_X_train, y_train)
```

(continues on next page)

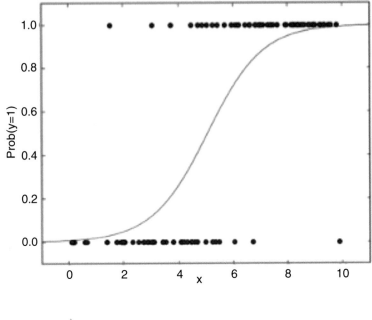

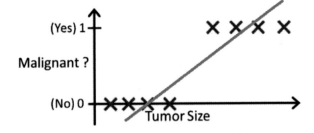

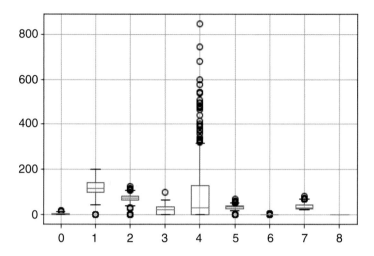

(continued from previous page)

```
y_pred_train1 = linear_classifier.predict(scaled_X_train)
cm1_train = confusion_matrix(y_train,y_pred_train1)
print("Regression")
print("==============================")
print(cm1_train)
acc_train1 = (cm1_train[0,0] + cm1_train[1,1]) / sum(sum(cm1_
↪train))
print("Regression TrainSet: Accurarcy %.2f%%" % (acc_
↪train1*100))
print("==============================")
y_pred1 = linear_classifier.predict(scaled_X_test)
cm1 = confusion_matrix(y_test,y_pred1)
print(cm1)
acc1 = (cm1[0,0] + cm1[1,1]) / sum(sum(cm1))
print("Regression Testset: Accurarcy %.2f%%" % (acc1*100))
print("==============================")
```

```
Regression
==============================
[[274   31]
 [ 62   93]]
Regression TrainSet: Accurarcy 79.78%
==============================
[[172   23]
 [ 53   60]]
Regression Testset: Accurarcy 75.32%
==============================
```

Sample result:

```
================================
[[274   31]
 [ 62   93]]
Regression TrainSet: Accurarcy 79.78%
================================
```

```
[[274   31]
 [ 62   93]]
```

Here is an example of a confusion matrix in python.

- 274 is when the Actual class is True and Predicted class is True.
- 31 is when the Actual class is True and Predicted class is False.
- 62 is when the Actual class is False and Predicted class is True.
- 93 is when the Actual class is False and Predicted class is False.

Improvement to Our Code

Recall we have written a helper function to help us to capture the logic of training the model, predicting the output and printing the train and test accuracy as well as confusion matrix? Let us put it to use here!

```
train_and_predict_using_model('Logistic Regression', linear_
↪classifier)
```

```
Logistic Regression
================================
Training confusion matrix:
[[274   31]
 [ 62   93]]
TrainSet: Accurarcy 79.78%
================================
[[172   23]
 [ 53   60]]
Testset: Accurarcy 75.32%
================================
```

We have managed to reduce multiple lines of code to a succinct function call. This is a huge improvement in terms of code maintenance and code changes. If we need to change any of our code, we only have to apply it on our `train_and_predict_using_model` function.

11.2   Decision Tree and Random Forest

The code and intuition behind Decision Tree and Random Forest is similar to that in regression. Thus, we will not be delving deeper into both models.

The code is as follows:

```
decision_tree_clf = tree.DecisionTreeClassifier()
train_and_predict_using_model('Decision Tree Classifier',␣
↪linear_classifier)

print(
    '\n\n'
)

rf_clf = RandomForestClassifier(n_estimators=100, max_depth=2,
↪random_state=0)
train_and_predict_using_model('Random Forest Classifier', rf_
↪clf)
```

```
Decision Tree Classifier
================================
Training confusion matrix:
[[274   31]
 [ 62   93]]
TrainSet: Accurarcy 79.78%
================================
[[172   23]
 [ 53   60]]
Testset: Accurarcy 75.32%
================================


Random Forest Classifier
================================
Training confusion matrix:
[[290   15]
 [ 96   59]]
TrainSet: Accurarcy 75.87%
================================
[[184   11]
 [ 80   33]]
Testset: Accurarcy 70.45%
================================
```

11.3 Neural Network

Lastly, we have neural network. Similar to logistic regression, we have to map our output from -inf to inf to 0 to 1. Here, we will have to add a Dense layer with a sigmoid activation function. For multi-class, we should use a softmax activation function.

```
model.add(Dense(1, activation='sigmoid'))
```

Here, we added a last layer mapping to a sigmoid function. Notice that we have 1 neuron in this layer as we would like to have 1 prediction. This might be different for multi-class, and we should always check out the documentation.

```
model.compile(loss='binary_crossentropy', optimizer='Adamax',
→metrics=['accuracy'])
```

Also, we would need to tell the model that we need to use a different loss function. Here, for binary classification problem (Yes/No). `binary_crossentropy` is the way to go. For multi-class classification problem, we might need to use `categorical_crossentropy` as the loss function.

```
#Neural network
#https://www.tensorflow.org/guide/keras/train_and_evaluate
model = Sequential()
model.add(Dense(5, input_dim=Y_position, activation='relu'))
model.add(Dense(1, activation='sigmoid'))

# Compile model
# https://www.tensorflow.org/guide/keras/train_and_evaluate
model.compile(loss='binary_crossentropy', optimizer='Adamax',
→metrics=['accuracy'])

# Fit the model
model.fit(scaled_X_train, y_train, epochs=1, batch_size=20,
→verbose=0)

# evaluate the model
scores = model.evaluate(scaled_X_train, y_train)

print("Neural Network Trainset: \n%s: %.2f%%" % (model.metrics_
→names[1], scores[1]*100))

predictions = model.predict(scaled_X_test)

y_pred = (predictions > 0.5)
y_pred = y_pred*1 #convert to 0,1 instead of True False
cm = confusion_matrix(y_test, y_pred)
print("=================================")
print("=================================")
print("Neural Network on testset confusion matrix")
print(cm)

## Get accurary from Confusion matrix
## Position 0,0 and 1,1 are the correct predictions
acc = (cm[0,0] + cm[1,1]) / sum(sum(cm))
print("Neural Network on TestSet: Accuracy %.2f%%" % (acc*100))
```

```
460/460 [==============================] - 0s 63us/step
Neural Network Trainset:
accuracy: 71.09%
=================================
=================================
Neural Network on testset confusion matrix
[[177  18]
 [ 78  35]]
Neural Network on TestSet: Accuracy 68.83%
```

From above, notice that the training accuracy is at 71%, which might be a case of underfitting. To improve our model, we can always increase the number of neurons/layer or increase the epoch for training.

```
#Neural network
#https://www.tensorflow.org/guide/keras/train_and_evaluate
model = Sequential()
model.add(Dense(10, input_dim=Y_position, activation='relu'))
model.add(Dense(256, activation='relu'))
model.add(Dropout(0.1))
model.add(Dense(256, activation='tanh'))
model.add(Dropout(0.1))
model.add(Dense(1, activation='sigmoid'))

# Compile model
# https://www.tensorflow.org/guide/keras/train_and_evaluate
model.compile(loss='binary_crossentropy', optimizer='RMSprop', ␣
↪metrics=['accuracy'])

# Fit the model
model.fit(scaled_X_train, y_train, epochs=200, batch_size=20, ␣
↪verbose=0)

# evaluate the model
scores = model.evaluate(scaled_X_train, y_train)

print("Neural Network Trainset: \n%s: %.2f%%" % (model.metrics_
↪names[1], scores[1]*100))

predictions = model.predict(scaled_X_test)

y_pred = (predictions > 0.5)
y_pred = y_pred*1 #convert to 0,1 instead of True False
cm = confusion_matrix(y_test, y_pred)
print("=================================")
print("=================================")
print("Neural Network on testset confusion matrix")
print(cm)

## Get accurary from Confusion matrix
## Position 0,0 and 1,1 are the correct predictions
```

(continues on next page)

(continued from previous page)

```
acc = (cm[0,0] + cm[1,1]) / sum(sum(cm))
print("Neural Network on TestSet: Accuracy %.2f%%" % (acc*100))
```

```
460/460 [==============================] - 0s 126us/step
Neural Network Trainset:
accuracy: 99.57%
=================================
=================================
Neural Network on testset confusion matrix
[[152  43]
 [ 43  70]]
Neural Network on TestSet: Accuracy 72.08%
```

Now, our accuracy on training has reached 99%. However, accuracy of test is still lower. This might be because of testing dataset differing from training dataset or overfitting. For overfitting, we will look at some regularization techniques. For now, adding `Dropout` layer and reducing training `epoch` would work just fine.

11.4   Logistic Regression

- https://scikit-learn.org/stable/modules/generated/sklearn.linear_model.
  LogisticRegression.html

```
class sklearn.linear_model.LogisticRegression(penalty='l2', *,
↪dual=False, tol=0.0001, C=1.0, fit_intercept=True,
intercept_scaling=1, class_weight=None, random_state=None,
↪solver='lbfgs', max_iter=100, multi_class='auto',
verbose=0, warm_start=False, n_jobs=None, l1_ratio=None)
```

11.5   Decision Tree

- https://scikit-learn.org/stable/modules/generated/sklearn.tree.
  DecisionTreeClassifier.html

```
class sklearn.tree.DecisionTreeClassifier(*, criterion='gini',
↪splitter='best', max_depth=None, min_samples_split=2,
min_samples_leaf=1, min_weight_fraction_leaf=0.0, max_
↪features=None, random_state=None, max_leaf_nodes=None,
min_impurity_decrease=0.0, min_impurity_split=None, class_
↪weight=None, presort='deprecated', ccp_alpha=0.0)
```

11.6   Feature Importance

```
RF = model3
importances = RF.feature_importances_
std = numpy.std([tree.feature_importances_ for tree in RF.
↪estimators_],
              axis=0)
indices = numpy.argsort(importances)[::-1]

# Print the feature ranking
print("Feature ranking:")

for f in range(X.shape[1]):
    print("%d. feature (Column index) %s (%f)" % (f + 1,␣
↪indices[f], importances[indices[f]]))
```

```
Feature ranking:
1. feature (Column index) 1 (0.307004)
2. feature (Column index) 7 (0.237150)
3. feature (Column index) 0 (0.129340)
4. feature (Column index) 5 (0.129255)
5. feature (Column index) 6 (0.069927)
6. feature (Column index) 4 (0.055137)
7. feature (Column index) 2 (0.044458)
8. feature (Column index) 3 (0.027729)
```

```
import seaborn as sns
import pandas as pd
boxplot = pd.DataFrame(dataset).boxplot()
```

11.7   Remove Outlier

```
df = pd.DataFrame(dataset)
quantile = df[4].quantile(0.99)
df1 = df[df[4] < quantile]
df.shape, df1.shape
```

```
((768, 9), (760, 9))
```

```
df1 = df1.dropna()
```

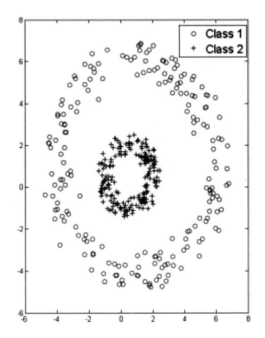

11.8   Use Top 3 Features

```
indices_top3 = indices[:3]
print(indices_top3)

# fix random seed for reproducibility
numpy.random.seed(7)
# load pima indians dataset
dataset = numpy.loadtxt("Diabetes (Edited).csv", delimiter=",")

df = pd.DataFrame(dataset)

Y_position = 8
TOP_N_FEATURE = 3

X = dataset[:,indices_top3]
Y = dataset[:,Y_position]
# create model
X_train, X_test, y_train, y_test = train_test_split(X, Y, test_
↪size=0.20, random_state=2020)


#scaling to around -2 to 2 (Z)
scaler = preprocessing.StandardScaler().fit(X_train)
scaled_X_train = scaler.transform(X_train)
```

(continues on next page)

(continued from previous page)

```
scaled_X_test = scaler.transform(X_test)

#Model 1 : linear regression
#https://scikit-learn.org/stable/modules/generated/sklearn.
↪linear_model.LogisticRegression.html
#class sklearn.linear_model.LogisticRegression(penalty='l2', *,
↪ dual=False, tol=0.0001, C=1.0, fit_intercept=True,
#intercept_scaling=1, class_weight=None, random_state=None, ␣
↪solver='lbfgs', max_iter=100, multi_class='auto',
#verbose=0, warm_start=False, n_jobs=None, l1_ratio=None)

linear_classifier = linear_model.LogisticRegression(random_
↪state=123)
linear_classifier.fit(scaled_X_train, y_train)
y_pred_train1 = linear_classifier.predict(scaled_X_train)
cm1_train = confusion_matrix(y_train,y_pred_train1)
print("Regression")
print("==============================")
print(cm1_train)
acc_train1 = (cm1_train[0,0] + cm1_train[1,1]) / sum(sum(cm1_
↪train))
print("Regression TrainSet: Accurarcy %.2f%%" % (acc_
↪train1*100))
print("==============================")
y_pred1 = linear_classifier.predict(scaled_X_test)
cm1 = confusion_matrix(y_test,y_pred1)
print(cm1)
acc1 = (cm1[0,0] + cm1[1,1]) / sum(sum(cm1))
print("Regression Testset: Accurarcy %.2f%%" % (acc1*100))
print("==============================")
print("==============================")
print("==============================")


#Model 2: decision tree
#https://scikit-learn.org/stable/modules/generated/sklearn.
↪tree.DecisionTreeClassifier.html
#class sklearn.tree.DecisionTreeClassifier(*, criterion='gini',
↪ splitter='best', max_depth=None, min_samples_split=2,
#min_samples_leaf=1, min_weight_fraction_leaf=0.0, max_
↪features=None, random_state=None, max_leaf_nodes=None,
#min_impurity_decrease=0.0, min_impurity_split=None, class_
↪weight=None, presort='deprecated', ccp_alpha=0.0)

clf = tree.DecisionTreeClassifier()
clf = clf.fit(scaled_X_train, y_train)
y_pred_train2 = clf.predict(scaled_X_train)
cm2_train = confusion_matrix(y_train,y_pred_train2)
print("Decision Tree")
print("==============================")
print(cm2_train)
```

(continues on next page)

(continued from previous page)

```
acc_train2 = (cm2_train[0,0] + cm2_train[1,1]) / sum(sum(cm2_
↪train))
print("Decsion Tree TrainSet: Accurarcy %.2f%%" % (acc_
↪train2*100))
print("===============================")
y_pred2 = clf.predict(scaled_X_test)
cm2 = confusion_matrix(y_test,y_pred2)
acc2 = (cm2[0,0] + cm2[1,1]) / sum(sum(cm2))
print(cm2)
print("Decision Tree Testset: Accurarcy %.2f%%" % (acc2*100))
print("===============================")
print("===============================")
print("===============================")


#Model 3 random forest
#https://scikit-learn.org/stable/modules/generated/sklearn.
↪ensemble.RandomForestClassifier.html
#class sklearn.ensemble.RandomForestClassifier(n_
↪estimators=100, *, criterion='gini', max_depth=None,
#min_samples_split=2, min_samples_leaf=1, min_weight_fraction_
↪leaf=0.0, max_features='auto',
#max_leaf_nodes=None, min_impurity_decrease=0.0, min_impurity_
↪split=None, bootstrap=True, oob_score=False,
#n_jobs=None, random_state=None, verbose=0, warm_start=False, ␣
↪class_weight=None, ccp_alpha=0.0, max_samples=None)[source]

model3 = RandomForestClassifier(n_estimators=100, max_depth=2,
↪random_state=0)
model3.fit(scaled_X_train, y_train)
y_predicted3 = model3.predict(scaled_X_test)

y_pred_train3 = model3.predict(scaled_X_train)
cm3_train = confusion_matrix(y_train,y_pred_train3)
print("Random Forest")
print("===============================")
print(cm3_train)
acc_train3 = (cm3_train[0,0] + cm3_train[1,1]) / sum(sum(cm3_
↪train))
print("Random Forest TrainSet: Accurarcy %.2f%%" % (acc_
↪train3*100))
print("===============================")
y_pred3 = model3.predict(scaled_X_test)
cm_test3 = confusion_matrix(y_test,y_pred3)
print(cm_test3)
acc_test3 = (cm_test3[0,0] + cm_test3[1,1]) / sum(sum(cm_
↪test3))
print("Random Forest Testset: Accurarcy %.2f%%" % (acc_
↪test3*100))
print("===============================")
print("===============================")
```

(continues on next page)

(continued from previous page)

```
print("================================")

#Model 4: XGBoost

print("Xgboost")
print("================================")
#class sklearn.ensemble.GradientBoostingClassifier(*, loss=
↪'deviance', learning_rate=0.1, n_estimators=100,
#subsample=1.0, criterion='friedman_mse', min_samples_split=2,␣
↪min_samples_leaf=1, min_weight_fraction_leaf=0.0,
#max_depth=3, min_impurity_decrease=0.0, min_impurity_
↪split=None, init=None, random_state=None, max_features=None,
#verbose=0, max_leaf_nodes=None, warm_start=False, presort=
↪'deprecated', validation_fraction=0.1,
#n_iter_no_change=None, tol=0.0001, ccp_alpha=0.0)[source]
#https://scikit-learn.org/stable/modules/generated/sklearn.
↪ensemble.GradientBoostingClassifier.html

model4 = GradientBoostingClassifier(random_state=0)
model4.fit(scaled_X_train, y_train)
y_pred_train4 = model4.predict(scaled_X_train)
cm4_train = confusion_matrix(y_train,y_pred_train4)
print(cm4_train)
acc_train4 = (cm4_train[0,0] + cm4_train[1,1]) / sum(sum(cm4_
↪train))
print("Xgboost TrainSet: Accurarcy %.2f%%" % (acc_train4*100))
predictions = model4.predict(scaled_X_test)
y_pred4 = (predictions > 0.5)
y_pred4 =y_pred4*1 #convert to 0,1 instead of True False
cm4 = confusion_matrix(y_test, y_pred4)
print("================================")
print("Xgboost on testset confusion matrix")
print(cm4)
acc4 = (cm4[0,0] + cm4[1,1]) / sum(sum(cm4))
print("Xgboost on TestSet: Accuracy %.2f%%" % (acc4*100))
print("================================")

#Model 5: neural network
#https://www.tensorflow.org/guide/keras/train_and_evaluate

model = Sequential()
model.add(Dense(10, input_dim=TOP_N_FEATURE, activation='relu
↪'))
#model.add(Dense(10, activation='relu'))
#model.add(Dropout(0.2))
model.add(Dense(1, activation='sigmoid'))
# Compile mode
# https://www.tensorflow.org/guide/keras/train_and_evaluate

model.compile(loss='binary_crossentropy', optimizer='Adamax',␣
↪metrics=['accuracy'])
```

(continues on next page)

(continued from previous page)

```
# Fit the model
model.fit(X_train, y_train, epochs=100, batch_size=5, ⏎
↪verbose=0)
# evaluate the model
scores = model.evaluate(X_train, y_train)
#print(scores)
print("Neural Network Trainset: \n%s: %.2f%%" % (model.metrics_
↪names[1], scores[1]*100))

predictions5 = model.predict(X_test)
#print(predictions)
#print('predictions shape:', predictions.shape)

y_pred5 = (predictions5 > 0.5)
y_pred5 = y_pred5*1 #convert to 0,1 instead of True False
cm5 = confusion_matrix(y_test, y_pred5)
print("=================================")
print("=================================")
print("Neural Network on testset confusion matrix")
print(cm5)

## Get accurary from Confusion matrix
## Position 0,0 and 1,1 are the correct predictions
acc5 = (cm5[0,0] + cm5[1,1]) / sum(sum(cm5))
print("Neural Network on TestSet: Accuracy %.2f%%" % ⏎
↪(acc5*100))
```

```
[1 7 0]
Regression
=================================
[[361  46]
 [105 102]]
Regression TrainSet: Accurarcy 75.41%
=================================
[[82 11]
 [30 31]]
Regression Testset: Accurarcy 73.38%
=================================
=================================
=================================
Decision Tree
=================================
[[407   0]
 [  0 207]]
Decsion Tree TrainSet: Accurarcy 100.00%
=================================
[[68 25]
 [32 29]]
Decision Tree Testset: Accurarcy 62.99%
=================================
=================================
```

(continues on next page)

(continued from previous page)

```
================================
Random Forest
================================
[[377  30]
 [128  79]]
Random Forest TrainSet: Accurarcy 74.27%
================================
[[87  6]
 [40 21]]
Random Forest Testset: Accurarcy 70.13%
================================
================================
================================
Xgboost
================================
[[389  18]
 [ 58 149]]
Xgboost TrainSet: Accurarcy 87.62%
================================
Xgboost on testset confusion matrix
[[80 13]
 [29 32]]
Xgboost on TestSet: Accuracy 72.73%
================================
20/20 [==============================] - 0s 1ms/step - loss: 0.
↪5480 - accuracy: 0.7671
Neural Network Trainset:
accuracy: 76.71%
================================
================================
Neural Network on testset confusion matrix
[[81 12]
 [29 32]]
Neural Network on TestSet: Accuracy 73.38%
```

11.9 SVM

```
from sklearn import svm

clf = svm.SVC()
train_and_predict_using_model("SVM (Classifier)", clf)
```

```
SVM (Classifier)
================================
Training confusion matrix:
[[361  46]
```

(continues on next page)

(continued from previous page)

```
   [101 106]]
TrainSet: Accurarcy 76.06%
==================================
[[84    9]
   [31 30]]
Testset: Accurarcy 74.03%
==================================
```

11.9.1   Important Hyper Parameters

For Support Vector Machines (SVM) here are some important parameters to take note of:

Kernel

Kernel Function generally transforms the training set of data so that a non-linear decision surface is able to transformed to a linear equation in a higher number of dimension spaces. Some of the possible parameters are as follows:

- Radial basis function
- Polynomial
- Sigmoid

Here is an illustrated use of a radial basis function (rbf) kernel.

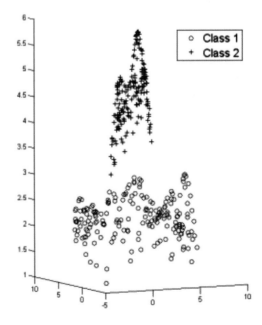

Another important parameter would be class_weight. Here, it is mainly used for unbalanced dataset.

```
rbf_svc = svm.SVC(kernel='rbf')
train_and_predict_using_model("SVM (RBF kernel)", rbf_svc)
```

```
SVM (RBF kernel)
================================
Training confusion matrix:
[[361  46]
 [101 106]]
TrainSet: Accurarcy 76.06%
================================
[[84   9]
 [31 30]]
Testset: Accurarcy 74.03%
================================
```

```
rbf_svc = svm.SVC(kernel='poly')
train_and_predict_using_model("SVM (polynomial kernel)", rbf_
↪svc)
```

```
SVM (polynomial kernel)
================================
Training confusion matrix:
[[393  14]
 [148  59]]
TrainSet: Accurarcy 73.62%
================================
[[89  4]
 [47 14]]
Testset: Accurarcy 66.88%
================================
```

```
rbf_svc = svm.SVC(kernel='sigmoid')
train_and_predict_using_model("SVM (sigmoid kernel)", rbf_svc)
```

```
SVM (sigmoid kernel)
================================
Training confusion matrix:
[[320  87]
 [112  95]]
TrainSet: Accurarcy 67.59%
================================
[[68 25]
 [35 26]]
Testset: Accurarcy 61.04%
================================
```

```
# fit the model and get the separating hyperplane using␣
↪weighted classes
wclf = svm.SVC(kernel='linear', class_weight={1:2})
train_and_predict_using_model('SVM uneven class weight', wclf)
```

```
SVM uneven class weight
================================
Training confusion matrix:
[[316  91]
 [ 75 132]]
TrainSet: Accurarcy 72.96%
================================
[[71 22]
 [18 43]]
Testset: Accurarcy 74.03%
================================
```

11.10   Naive Bayes

It is a classification technique based on Bayes' Theorem with an assumption of independence among predictors. In simple terms, a Naive Bayes classifier assumes that the presence of a particular feature in a class is unrelated to the presence of any other feature.

For example, a fruit may be considered to be an apple if it is red, round, and about 3 inches in diameter. Even if these features depend on each other or upon the existence of the other features, all of these properties independently contribute to the probability that this fruit is an apple and that is why it is known as "Naive."

Naive Bayes model is easy to build and particularly useful for very large datasets. Along with simplicity, Naive Bayes is known to outperform even highly sophisticated classification methods.

Bayes' theorem provides a way of calculating posterior probability $P(c|x)$ from $P(c)$, $P(x)$, and $P(x|c)$

$$P(A \mid B) = \frac{P(B \mid A) \cdot P(A)}{P(B)}$$

A, B = events

$P(A|B)$ = probability of A given B is true

$P(B|A)$ = probability of B given A is true

$P(A), P(B)$ = the independent probabilities of A and B

```
from sklearn.naive_bayes import GaussianNB

# maximum likelihood

gnb = GaussianNB()
train_and_predict_using_model("Naive Bayes", gnb)
```

```
Naive Bayes
================================
Training confusion matrix:
[[337  70]
 [ 93 114]]
TrainSet: Accurarcy 73.45%
================================
[[78 15]
 [28 33]]
Testset: Accurarcy 72.08%
================================
```

```
import numpy as np
from sklearn.datasets import make_classification
from sklearn.naive_bayes import GaussianNB

X, y = make_classification(n_samples=1000, weights=[0.1, 0.9])
# your GNB estimator
gnb = GaussianNB()
gnb.fit(X, y)

print("model prior {} close to your defined prior of {}".
→format(gnb.class_prior_, [0.1,0.9]))
```

```
model prior [0.105 0.895] close to your defined prior of [0.1,␣
→0.9]
```

11.11 Sample Code

```
import pandas as pd

df=pd.read_csv("C:/Users/User/Dropbox/TT Library/AI Model/
→Python/Treynor (Classification).csv")

print(df)

df=df.dropna()
print(df)
```

(continues on next page)

(continued from previous page)

```
for i in df.columns:
    df=df[pd.to_numeric(df[i], errors='coerce').notnull()]
↪#make it to null then remove null
print(df)
```

```
import seaborn as sns

sns.barplot(x="Class", y="size_type", data=df)
```

```
import matplotlib.pyplot as plt

df.hist()
plt.show()
```

```
sns.barplot(x="Class", y= "Blend", data=df)
```

```
import numpy as np
from scipy import stats

print(df)
z_scores = stats.zscore(df.astype(np.float))
print(z_scores)
abs_z_scores = np.abs(z_scores)
filtered_entries = (abs_z_scores < 3).all(axis=1)
print(filtered_entries)
df = df[filtered_entries]

print(df)
```

```
df.describe()
```

```
df.corr()
```

```
import seaborn as sns

sns.heatmap(df.corr())
```

```
#Split X and Y

X=df.iloc[:,0:len(df.columns)-1]
print(X)
Y=df.iloc[:,len(df.columns)-1]
print(Y)
```

```
dummy=pd.get_dummies(X["Blend"])
dummy.head()
```

```
X=X.merge(dummy, left_index=True, right_index=True)
X.head()
```

```
X=X.drop("Blend", axis="columns")
X.head()
```

```
#Normalization

X["return_rating"]=stats.zscore(X["return_rating"].astype(np.
↪float))

print(X)
```

```
#split train test

from sklearn.model_selection import train_test_split

X_train, X_test, Y_train, Y_test = train_test_split(X, Y, test_
↪size=0.3)
print(X_train)
print(X_test)
print(Y_train)
print(Y_test)
```

```
from sklearn import linear_model
from sklearn.metrics import confusion_matrix

model = linear_model.LogisticRegression(max_iter=1000)
model.fit(X_train, Y_train)
pred=model.predict(X_train)
cm=confusion_matrix(pred, Y_train)
print(cm)
```

```
accuracy=(cm[0,0]+cm[1,1])/sum(sum(cm))
print(accuracy)
```

```
pred=model.predict(X_test)
cm=confusion_matrix(pred, Y_test)
print(cm)
```

```
accuracy=(cm[0,0]+cm[1,1])/sum(sum(cm))
print(accuracy)
```

```
import statsmodels.api as sm
logit_model=sm.Logit(Y,X)
result=logit_model.fit()
print(result.summary2())
```

```python
from sklearn.tree import DecisionTreeClassifier

model = DecisionTreeClassifier()
model.fit(X_train, Y_train)
pred=model.predict(X_train)
cm=confusion_matrix(pred, Y_train)
print(cm)
accuracy=(cm[0,0]+cm[1,1])/sum(sum(cm))
print(accuracy)
```

```python
pred=model.predict(X_test)
cm=confusion_matrix(pred, Y_test)
print(cm)
accuracy=(cm[0,0]+cm[1,1])/sum(sum(cm))
print(accuracy)
```

```python
# random forest
from sklearn.ensemble import RandomForestClassifier
model=RandomForestClassifier()
model.fit(X_train,Y_train)
Y_predict=model.predict(X_train)
cm=confusion_matrix(Y_train, Y_predict)
print(cm)

accuracy=(cm[0,0]+cm[1,1])/sum(sum(cm))
print(accuracy)
pred=model.predict(X_test)
cm=confusion_matrix(pred, Y_test)
print(cm)
accuracy=(cm[0,0]+cm[1,1])/sum(sum(cm))
print(accuracy)
```

```python
from sklearn.ensemble import GradientBoostingClassifier

model=GradientBoostingClassifier()
model.fit(X_train, Y_train)
pred = model.predict(X_train)
cm=confusion_matrix(Y_train, pred)
print(cm)

accuracy=(cm[0,0]+cm[1,1])/sum(sum(cm))
print(accuracy)
pred=model.predict(X_test)
cm=confusion_matrix(Y_test, pred)
print(cm)

accuracy=(cm[0,0]+cm[1,1])/sum(sum(cm))
print(accuracy)
```

```
from keras.models import Sequential
from keras.layers import Dense, Dropout

model=Sequential()
model.add(Dense(10, input_dim=len(X_train.columns), activation=
↪'relu'))
model.add(Dropout(0.2))
model.add(Dense(10, activation="relu"))
model.add(Dropout(0.2))
model.add(Dense(1, activation='sigmoid'))

model.compile(loss = 'binary_crossentropy', optimizer = 'Adamax
↪', metrics = ['accuracy'])

model.fit(X_train, Y_train, epochs=100, batch_size=10,␣
↪verbose=0)

score=model.evaluate(X_train, Y_train)
print(score[1])
score=model.evaluate(X_test, Y_test)
print(score[1])

pred=model.predict(X_train)
pred=np.where(pred>0.5,1,0)
cm=confusion_matrix(pred, Y_train)
print(cm)
accuracy=(cm[0,0]+cm[1,1])/sum(sum(cm))
print(accuracy)

pred=model.predict(X_test)
pred=np.where(pred>0.5,1,0)
cm=confusion_matrix(pred, Y_test)
print(cm)
accuracy=(cm[0,0]+cm[1,1])/sum(sum(cm))
print(accuracy)
```

Chapter 12
Clustering

Abstract Supervised algorithms use labeled data as an input for developing a prediction model. However, the amount of unlabeled data collected often far exceeds that of labeled data. Unsupervised algorithms, such as clustering algorithms, are algorithms that are able to make use of these unlabeled data to extract useful insights. One example is the K-means algorithm which clusters the data into K different clusters. The elbow method will be used to select a suitable number of clusters to select the value of K.

Learning outcomes:

- Understand the difference between supervised and unsupervised algorithms.
- Learn and apply the K-means algorithm for clustering tasks using sklearn.
- Learn the elbow method to select a suitable number of clusters.

12.1 What Is Clustering?

Clustering is the task of dividing the population or data points into a number of groups, such that data points in the same groups are more similar to other data points within the group and dissimilar to the data points in other groups. Clustering is a form of unsupervised algorithm. This means that unlike classification or regression, clustering does not require ground truth labeled data. Such algorithms are capable of finding groups that are not explicitly labeled and identify underlying patterns that might appear in the dataset. One of the simplest, yet effective clustering algorithm is the K-means algorithm.

T. T. Teoh, Z. Rong, *Artificial Intelligence with Python*,
Machine Learning: Foundations, Methodologies, and Applications,
https://doi.org/10.1007/978-981-16-8615-3_12

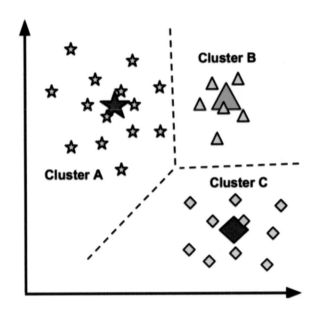

12.2   K-Means

K-means is used for a variety of cases, such as:

• Customer profiling
• Market segmentation
• Computer vision
• Geo-statistics
• Astronomy

The K-means algorithm clusters data by trying to separate samples in n groups of equal variance, minimizing a criterion known as the inertia or within-cluster sum-of-squares. The K-means algorithm aims to choose centroid that minimizes the inertia or within-cluster sum-of-squares criterion:

$$\sum_{i=0}^{n} \min_{\mu_j \in C}(||x_i - \mu_j||^2)$$

The steps for the K-means algorithm are as follows:

1. Ask user how many clusters they would like (e.g., k=5).
2. Randomly guess k cluster center locations.

3. Each data point identifies which center is closest to according to the sum-of-squares criterion. (Thus, each center "owns" a set of data points.)
4. Reposition the k cluster center locations by minimizing the sum-of-squares criterion. This can be achieved by setting the new locations as the average of all the points in a cluster.
5. Repeat steps 3 and 4 until no new data points are added or removed from all clusters or the predefined maximum number of iterations has been reached.

12.3   The Elbow Method

As you can see in the first step of the K-means algorithm, the user has to specify the number of clusters to be used for the algorithm. We can do this by attempting the K-means for various values of K and visually selecting the K-value using the elbow method. We would like a small sum-of-squares error, and however, the sum-of-squares error tends to decrease toward 0 as we increase the value of k. Sum-of-squares will decrease toward 0 with increasing k, because when k is equal to the number of data points, each data point is its own cluster, and there will be no error between it and the center of its cluster.

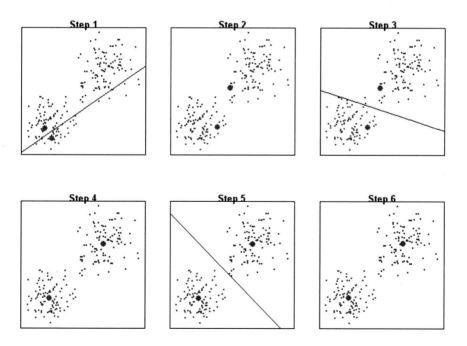

Elbow method

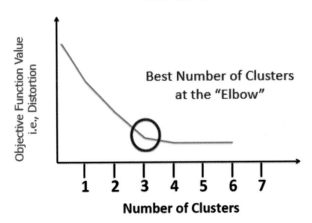

The following code example shows the K-means algorithm and the elbow visualization using the Iris dataset which can be obtained from: https://www.kaggle.com/uciml/iris:

```python
import numpy as np
import pandas as pd

df = pd.read_csv("iris.csv")
print(df)
df["Species"].unique()
df = df.replace("Iris-setosa", 0)
df=df.replace("Iris-versicolor", 1)
df = df.replace("Iris-virginica", 2)

X=df.loc[:, ["SepalLengthCm","SepalWidthCm","PetalLengthCm",
  "PetalWidthCm"]]
Y=df['Species']
print(X)
print(Y)

from sklearn.cluster import KMeans
model=KMeans(n_clusters=3, random_state=2021)
model.fit(X,Y)
pred=model.predict(X)

from sklearn.metrics import confusion_matrix
cm=confusion_matrix(pred, Y)
print(cm)

accuracy=(cm[0,0]+cm[1,1]+cm[2,2])/sum(sum(cm))  #cm[rows,
  columns]
print(accuracy)
```

(continues on next page)

(continued from previous page)

```
from yellowbrick.cluster import KElbowVisualizer

visualizer = KElbowVisualizer(model, k=(2,15))

visualizer.fit(X)
visualizer.show()
```

```
        SepalLength  SepalWidth  PetalLength  PetalWidth          ↵
  ↪   Iris
0            5.1         3.5          1.4         0.2       Iris-
↪setosa
1            4.9         3.0          1.4         0.2       Iris-
↪setosa
2            4.7         3.2          1.3         0.2       Iris-
↪setosa
3            4.6         3.1          1.5         0.2       Iris-
↪setosa
4            5.0         3.6          1.4         0.2       Iris-
↪setosa
..           ...         ...          ...         ...              ↵
  ↪   ...
145          6.7         3.0          5.2         2.3   Iris-
↪virginica
146          6.3         2.5          5.0         1.9   Iris-
↪virginica
147          6.5         3.0          5.2         2.0   Iris-
↪virginica
148          6.2         3.4          5.4         2.3   Iris-
↪virginica
149          5.9         3.0          5.1         1.8   Iris-
↪virginica

[150 rows x 5 columns]
        SepalLength  SepalWidth  PetalLength  PetalWidth
0            5.1         3.5          1.4         0.2
1            4.9         3.0          1.4         0.2
2            4.7         3.2          1.3         0.2
3            4.6         3.1          1.5         0.2
4            5.0         3.6          1.4         0.2
..           ...         ...          ...         ...
145          6.7         3.0          5.2         2.3
146          6.3         2.5          5.0         1.9
147          6.5         3.0          5.2         2.0
148          6.2         3.4          5.4         2.3
149          5.9         3.0          5.1         1.8

[150 rows x 4 columns]
0        0
1        0
2        0
3        0
```

(continues on next page)

(continued from previous page)

```
4         0
          ..
145       2
146       2
147       2
148       2
149       2
Name: Iris, Length: 150, dtype: int64
[[50  0  0]
 [ 0 48 14]
 [ 0  2 36]]
0.8933333333333333
```

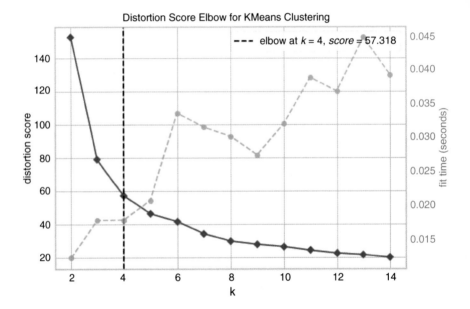

```
<AxesSubplot:title={'center':'Distortion Score Elbow for␣
↪KMeans Clustering'}, xlabel='k', ylabel='distortion score'>
```

Chapter 13
Association Rules

Abstract Association rule analysis is a technique which discovers the association between various items within large datasets in different types of databases and can be used as a form of feature engineering. The Apriori algorithm covered, mines for frequent itemsets and association rules in a database. Support, Lift, Conviction, and Confidence are important values that represent the strength of an association.

Learning outcomes:

- Learn the general concept of association rule mining.
- Understand concepts of support, lift, and confidence in a rules.
- Learn the Apriori algorithm for association rule mining.

13.1 What Are Association Rules

Association rules are "if-then" statements that help to show the probability of relationships between data items, within large datasets in various types of databases.

Association rule mining is the process of engineering data into a predictive feature in order to fit the requirements or to improve the performance of a model. The Apriori Algorithm is an algorithm used to perform association rule mining over a structured dataset.

Concepts in Association Rules:

- Transaction Record. A Transaction record is a record of all the purchases made in one transaction. For example, Receipt from NTUC supermarket, Invoice for mobile phone contract, Invoice of software purchase, etc.
- Transactional Data Binary Format:

  - Each row is a transaction record.
  - Each column represents a unique item.

T. T. Teoh, Z. Rong, *Artificial Intelligence with Python*,
Machine Learning: Foundations, Methodologies, and Applications,
https://doi.org/10.1007/978-981-16-8615-3_13

– Each row then records a sequence of 0 or 1 (the item in that column was purchased).

Note: Some software require the dataset to be in transactional data binary format in order to perform association analysis. Not SAS EM. An example of a Transactional Data Binary Format:

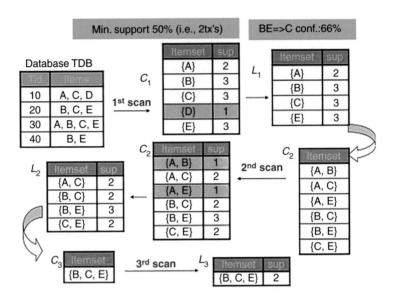

In the image above, each row (transaction ID) shows the items purchased in that single transaction (aka receipt).If there are many possible items, dataset is likely to be sparse, i.e. many zeros or NAs.

- Itemset. A Itemset is a set of items in a transaction. An itemset of size 3 means there are 3 items in the set. In general, it can be of any size, unless specified otherwise.
- Association Rule

  – Forms: X => Y
  – X associated with Y.
  – X is the "antecedent" itemset; Y is the "consequent" itemset.
  – There might be more than one item in X or Y.

13.2 Apriori Algorithm

Apriori Algorithm Steps

1. Candidate itemsets are generated using only the large itemsets of the previous pass without considering the transactions in the database.

2. The large itemset of the previous pass is joined with itself to generate all itemsets whose size is higher by 1.
3. Each generated itemset that has a subset which is not large is deleted. The remaining itemsets are the candidate ones.

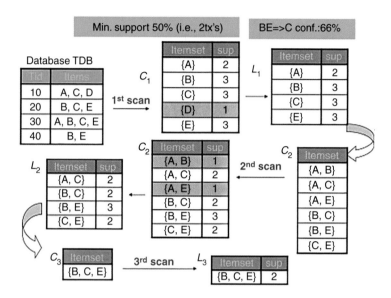

The Apriori algorithm assumes all subsets of a frequent itemset to be frequent. Similarly, for any infrequent itemset, all its supersets must also be infrequent.

13.3   Measures for Association Rules

In order to select the interesting rules out of multiple possible rules from this small business scenario, we will be using the following measures:

- Support
- Confidence
- Lift
- Conviction

1. Support is an indication of how frequently the item appears in the dataset. For example, how popular a product is in a shop. The support for the combination A and B would be,- P(AB) or P(A) for Individual A.
2. Confidence is an indication of how often the rule has been found to be true. It indicates how reliable the rule is. For example, how likely is it that someone would buy toothpaste when buying a toothbrush. In other words, confidence is the

conditional probability of the consequent given the antecedent,- P(B|A), where P(B|A) = P(AB)/P(A).

3. Lift is a metric to measure the ratio of the confidence of products occurring together if they were statistically independent. For example, how likely is another product purchased when purchasing a product, while controlling how popular the other product is.

A lift score that is close to 1 indicates that the antecedent and the consequent are independent and occurrence of antecedent has no impact on occurrence of consequent. A Lift score that is greater than 1 indicates that the antecedent and consequent are dependent to each other, and the occurrence of antecedent has a positive impact on occurrence of consequent. A lift score that is smaller than 1 indicates that the antecedent and the consequent are substitute each other that means the existence of antecedent has a negative impact to consequent or vice versa.

Consider an association rule "if A then B." The lift for the rule is defined as- P(B|A)/P(B), which is also P(AB)/(P(A)*P(B)).As shown in the formula, lift is symmetric in that the lift for "if A then B" is the same as the lift for "if B then A."

4. Conviction score is a ratio between the probability that one product occurs without another while they were dependent and the actual probability of one products' existence without another. It measures the implied strength of the rule from statistical independence. For example, if the (oranges) → (apples) association has a conviction score of 1.5; the rule would be incorrect 1.5 times more often (50% more often) if the association between the two were totally independent.

The Conviction score of A -> B would be defined as: $-(1-\text{Support}(B))/(1-\text{Confidence}(A \to B))$

By using the earlier shown dataset we can calculate the support, confidence, and lift of a rule.

For example, for the rule {milk, bread} => Butter, we can calculate the following measures:

- Support ({milk}) = 2/5 = 0.4
- Support ({milk, bread}) = 2/5 = 0.4
- Confidence ({milk, bread} => Butter)= 1/2 = 0.5
- Lift ({milk, bread} => Butter) = (1/2) / (2/5) = 1.25

We have generated a sample dataset consisting of nine transactions in an hour. Each transaction is a combination of 0s and 1s, where 0 represents the absence of an item and 1 represents the presence of it.

We can find multiple rules from this scenario. For example, in a transaction of milk and bread, if milk is bought, then customers also buy bread.

```
transactions  =   [["milk", "bread"],
                  ["butter"],
```

(continues on next page)

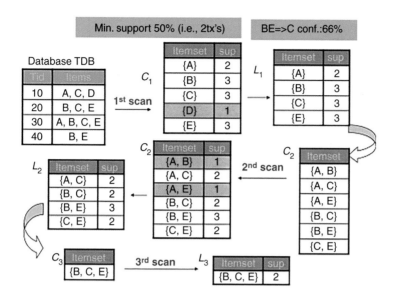

(continued from previous page)

```
                    ["beer", "diapers"],
                    ["milk", "bread", "butter"],
                    ["bread"],
                    ["beer"],
                    ["beer", "diapers"],
                    ["diapers"],
                    ["beer", "diapers"]]

import pandas as pd
from mlxtend.preprocessing import TransactionEncoder

te = TransactionEncoder()
te_ary = te.fit(transactions).transform(transactions)
df = pd.DataFrame(te_ary, columns=te.columns_).astype(int)
df
```

```
    beer   bread   butter   diapers   milk
0    0      1        0         0        1
1    0      0        1         0        0
2    1      0        0         1        0
3    0      1        1         0        1
4    0      1        0         0        0
5    1      0        0         0        0
6    1      0        0         1        0
7    0      0        0         1        0
8    1      0        0         1        0
```

```
item = apriori(df, use_colnames=True, min_support=0.2)
print(item)
print("=========")
rules1 = association_rules(item, metric = 'confidence', min_
↪threshold=0.7)
print(rules1)
```

```
     support            itemsets
0   0.444444             (beer)
1   0.333333            (bread)
2   0.222222           (butter)
3   0.444444          (diapers)
4   0.222222             (milk)
5   0.333333   (beer, diapers)
6   0.222222     (milk, bread)
=========
  antecedents consequents   antecedent support   consequent
↪support    support  \
0      (beer)    (diapers)             0.444444                0.
↪444444   0.333333
1   (diapers)       (beer)             0.444444                0.
↪444444   0.333333
2      (milk)      (bread)             0.222222                0.
↪333333   0.222222

   confidence    lift   leverage   conviction
0        0.75   1.6875   0.135802     2.222222
1        0.75   1.6875   0.135802     2.222222
2        1.00   3.0000   0.148148          inf
```

Part III
Artificial Intelligence Implementations

Chapter 14
Text Mining

Abstract Text mining is the process of extracting meaning from unstructured text documents using both machine learning and natural language processing techniques. It enables the process of converting reviews into specific recommendations that can be used. Text data would be represented in structured formats through converting text into numerical representations. Structured text data can then be ingested by Artificial Intelligence algorithms for various tasks such as sentence topic classification and keyword extraction.

Learning outcomes:

- Represent text data in structured and easy-to-consume formats for text mining.
- Perform sentence classification tasks on text data.
- Identify important keywords for sentence classification.

Text mining combines both machine learning and natural language processing (NLP) to draw meaning from unstructured text documents. Text mining is the driving force behind how a business analyst turns 50,000 hotel guest reviews into specific recommendations, how a workforce analyst improves productivity and reduces employee turnover, and how companies are automating processes using chatbots.

A very popular and current strategy in this field is Vectorized Term Frequency and Inverse Document Frequency (TF-IDF) representation. In fact, Google search engine also uses this technique when a word is searched. It is based on unsupervised learning technique. TF-IDF converts your document text into a bag of words and then assigns a weighted term to each word. In this chapter, we will discuss how to use text mining techniques to get meaningful results for text classification.

14.1 Read Data

```
import pandas as pd

#this assumes one json item per line in json file
df=pd.read_json("TFIDF_news.json", lines=True)
```

```
df.dtypes
```

```
short_description                object
headline                         object
date                     datetime64[ns]
link                             object
authors                          object
category                         object
dtype: object
```

```
#number of rows (datapoints)
len(df)
```

```
124989
```

```
# Take sample of 3 to view the data
df.sample(3)
```

```
                                  short_description  \
100659  The hardest battles are not fault in the stree...
74559   Mizzou seems to have catalyzed years of tensio...
48985                    But also hilariously difficult.

                                           headline    ⌴
↪ date  \
100659    American Sniper Dials in on the Reality of War 2015-
↪01-23
74559   Campus Racism Protests Didn't Come Out Of Nowh... 2015-
↪11-16
48985   These People Took On Puerto Rican Slang And It... 2016-
↪09-02

                                               link  \
100659  https://www.huffingtonpost.com/entry/american-...
74559   https://www.huffingtonpost.com/entry/campus-ra...
48985   https://www.huffingtonpost.com/entry/these-peo...

                                            authors    ⌴
↪ category
100659  Zachary Bell, Contributor United States Marine ... ⌴
↪ENTERTAINMENT
```

(continues on next page)

(continued from previous page)

```
74559     Tyler Kingkade, Lilly Workneh, and Ryan Grenoble    ␣
↪   COLLEGE
48985                                    Carolina Moreno  ␣
↪LATINO VOICES
```

14.2 Date Range

Articles are between July 2014 and July 2018.

```
df.date.hist(figsize=(12,6),color='#86bf91',)
```

```
<matplotlib.axes._subplots.AxesSubplot at 0x1a695a80508>
```

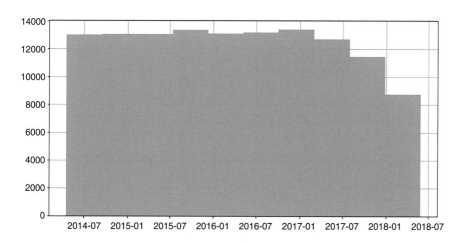

14.3 Category Distribution

In our data, there are a total of 31 categories.

```
len(set(df['category'].values))
```

```
31
```

Most of the articles are related to politics. Education related articles have the lowest volume.

```
import matplotlib
import numpy as np
cmap = matplotlib.cm.get_cmap('Spectral')
rgba = [cmap(i) for i in np.linspace(0,1,len(set(df['category
↪'].values)))]
df['category'].value_counts().plot(kind='bar',color =rgba)
```

```
<matplotlib.axes._subplots.AxesSubplot at 0x1a6942753c8>
```

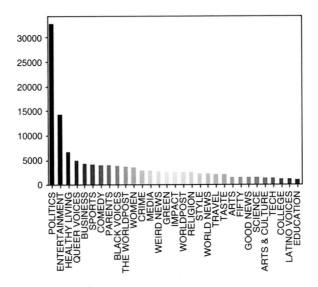

14.4   Texts for Classification

In our example, we will only use the headline to predict category. Also, we will only be using 2 categories, sports and crime, for simplicity. Notice that we are using CRIME and COMEDY categories from our dataset.

```
df_orig=df.copy()
df = df_orig[df_orig['category'].isin(['CRIME','COMEDY'])]
print(df.shape)
df.head()
df = df.loc[:, ['headline','category']]
df['category'].value_counts().plot(kind='bar',color =['r','b'])
```

```
(6864, 6)
```

```
<matplotlib.axes._subplots.AxesSubplot at 0x1a695c76388>
```

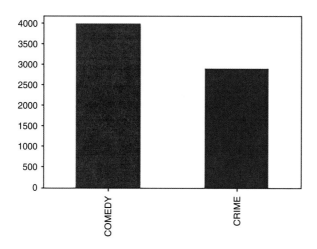

14.5   Vectorize

Text Vectorization is the process of converting text into numerical representation.

- Binary Term Frequency
- Bag of Words (BoW) Term Frequency
- (L1) Normalized Term Frequency
- (L2) Normalized TF-IDF
- Word2Vec

Binary Term Frequency
Binary Term Frequency captures presence (1) or absence (0) of term in document.

Bag of Words (BoW) Term Frequency
Bag of Words (BoW) Term Frequency captures frequency of a term in the document.
This is unlike Binary Term Frequency, which only captures whether a term is in the
document or is not in the document.

The following code is an example of Bag of Words Term Frequency:

```
from sklearn.feature_extraction.text import CountVectorizer

sample_doc = ["Hello I am a boy", "Hello I am a student", "My␣
↪name is Jill"]
```

(continues on next page)

(continued from previous page)

```
cv=CountVectorizer(max_df=0.85)
word_count_vector=cv.fit_transform(sample_doc)
word_count_vector_arr = word_count_vector.toarray()
pd.DataFrame(word_count_vector_arr,columns=sorted(cv.
  ↪vocabulary_, key=cv.vocabulary_.get))
```

	am	boy	hello	is	jill	my	name	student
0	1	1	1	0	0	0	0	0
1	1	0	1	0	0	0	0	1
2	0	0	0	1	1	1	1	0

An important note is the vocabulary is placed in a dictionary and python dictionaries are unsorted. Notice that the header in the following code is different from the first example.

```
## Wrong example
pd.DataFrame(word_count_vector_arr,columns=cv.vocabulary_)
```

	hello	am	boy	student	my	name	is	jill
0	1	1	1	0	0	0	0	0
1	1	0	1	0	0	0	0	1
2	0	0	0	1	1	1	1	0

This is because of dictionary in python. See below:

```
cv.vocabulary_
```

```
{'hello': 2,
 'am': 0,
 'boy': 1,
 'student': 7,
 'my': 5,
 'name': 6,
 'is': 3,
 'jill': 4}
```

14.6   CountVectorizer

Let's move on to our code example. Now, let's look at 10 words from our vocabulary. We have also removed words that appear in 95% of documents. In text analytics, such words (stop words) are not meaningful. An intuitive approach to understanding removal of stop words is that in a sentence, many words are present because of grammatical rules and do not add extra content or meaning. Ignoring such words would allow us to distill the key essence of a document and sentence. Sweet, after removing stop words by having maxdf=0.95, our key words are mostly crime and comedy related.

```
from sklearn.feature_extraction.text import CountVectorizer
docs=df['headline'].tolist()
# create a vocabulary of words,
# ignore words that appear in 85% of documents,
# eliminate stop words
cv=CountVectorizer(max_df=0.95)
word_count_vector=cv.fit_transform(docs)
list(cv.vocabulary_.keys())[:10]
```

```
['there',
 'were',
 'mass',
 'shootings',
 'in',
 'texas',
 'last',
 'week',
 'but',
 'only']
```

We can also use machine learning models learnt previously to classify our headlines! See code below:

```
from sklearn.feature_extraction.text import TfidfTransformer
from sklearn.model_selection import train_test_split
from sklearn.linear_model import LogisticRegression
from sklearn.metrics import confusion_matrix

df['category_is_crime'] = df['category']=='CRIME'
X_train, X_test, y_train, y_test = train_test_split(word_count_
↪vector, df['category_is_crime'], test_size=0.2, random_
↪state=42)
```

Wow, we achieve 95.19% in classifying headlines. This is a remarkable feat for our machine!

```
model1 = LogisticRegression()
model1.fit(X_train, y_train)

y_pred = model1.predict(X_test)
cm=confusion_matrix(y_test, y_pred)
print(cm)
acc=(cm[0,0]+cm[1,1])/sum(sum(cm))
print('Accuracy of a simple linear model with CountVectorizer␣
↪is .... {:.2f}%'.format(acc*100))
```

```
[[766  26]
 [ 40 541]]
Accuracy of a simple linear model with CountVectorizer is ....␣
↪95.19%
```

14.7 TF-IDF

We implement a L2 Normalized TF-IDF transformation on the Word counts provided, as shown below in this example.

```
tfidf_transformer=TfidfTransformer(smooth_idf=True,use_
↪idf=True)
tfidf_x_train = tfidf_transformer.fit_transform(X_train)
model1 = LogisticRegression()
model1.fit(tfidf_x_train, y_train)
tfidf_x_test = tfidf_transformer.transform(X_test)
y_pred = model1.predict(tfidf_x_test)
cm=confusion_matrix(y_test, y_pred)
print(cm)
acc=(cm[0,0]+cm[1,1])/sum(sum(cm))
print('Accuracy of a simple linear model with TFIDF is .... {:.
↪2f}%'.format(acc*100))
```

```
[[777   15]
 [ 57  524]]
Accuracy of a simple linear model with TFIDF is .... 94.76%
```

For L1 Normalized Term Frequency, it can be done by passing: $norm =' l1'$, $use_idf = False$, and $smooth_idf = False$ into TfidfTransformer.

14.8 Feature Extraction with TF-IDF

Apart from text classification, we can use TF-IDF to discover "important" keywords. Here is a few example that shows the importance of each individual word. Such technique is simple and easy to use. But on a cautionary note, using TF-IDF is heavily dependent on the input data and the importance of the text is closely related to the frequency in the document and across the entire data.

```
## Important keywords extraction using tfidf
print(df.iloc[1].headline)
vector = cv.transform([df.iloc[1].headline])
tfidf_vector = tfidf_transformer.transform(vector)
coo_matrix = tfidf_vector.tocoo()
tuples = zip(coo_matrix.col, coo_matrix.data)
sorted_tuple = sorted(tuples, key=lambda x: (x[1], x[0]),⎵
↪reverse=True)
[(cv.get_feature_names()[i[0]],i[1]) for i in sorted_tuple]
```

```
Rachel Dolezal Faces Felony Charges For Welfare Fraud
```

```
[('welfare', 0.413332601468908),
 ('felony', 0.413332601468908),
 ('dolezal', 0.413332601468908),
 ('rachel', 0.3885287853920158),
 ('fraud', 0.3599880238280249),
 ('faces', 0.3103803916742406),
 ('charges', 0.2954500640160872),
 ('for', 0.15262948420298186)]
```

```
## Important keywords extraction using tfidf
print(df.iloc[5].headline)
vector = cv.transform([df.iloc[5].headline])
tfidf_vector = tfidf_transformer.transform(vector)
coo_matrix = tfidf_vector.tocoo()
tuples = zip(coo_matrix.col, coo_matrix.data)
sorted_tuple = sorted(tuples, key=lambda x: (x[1], x[0]),␣
↪reverse=True)
[(cv.get_feature_names()[i[0]],i[1]) for i in sorted_tuple]
```

```
Man Faces Charges After Pulling Knife, Stun Gun On Muslim␣
↪Students At McDonald's
```

```
[('stun', 0.37604716794652987),
 ('pulling', 0.3658447343442784),
 ('knife', 0.32581708572483403),
 ('mcdonald', 0.32215742177499496),
 ('students', 0.30480662832662847),
 ('faces', 0.2922589939460096),
 ('muslim', 0.28707744879148683),
 ('charges', 0.27820036570239326),
 ('gun', 0.24718607863715278),
 ('at', 0.17925932409191916),
 ('after', 0.17428789091260877),
 ('man', 0.17199120825269787),
 ('on', 0.15323370190782204)]
```

```
comedy_1 = df[~df['category_is_crime']].iloc[0].headline
print(comedy_1)
```

```
Trump's New 'MAGA'-Themed Swimwear Sinks On Twitter
```

```
## Important keywords extraction using tfidf
vector = cv.transform([comedy_1])
tfidf_vector = tfidf_transformer.transform(vector)
coo_matrix = tfidf_vector.tocoo()
tuples = zip(coo_matrix.col, coo_matrix.data)
sorted_tuple = sorted(tuples, key=lambda x: (x[1], x[0]),␣
↪reverse=True)
[(cv.get_feature_names()[i[0]],i[1]) for i in sorted_tuple]
```

```
[('swimwear', 0.4735563110982704),
 ('sinks', 0.4735563110982704),
 ('maga', 0.4735563110982704),
 ('themed', 0.37841071080711314),
 ('twitter', 0.2770106227768904),
 ('new', 0.22822300865931006),
 ('on', 0.17796879475963143),
 ('trump', 0.15344404805174222)]
```

14.9   Sample Code

```python
import requests
from bs4 import BeautifulSoup

page = requests.get("http://www.facebook.com")
soup = BeautifulSoup(page.content, "html.parser")

print(soup)
```

```python
from nltk.tokenize import TweetTokenizer
tknzr = TweetTokenizer()
s0 = "This is a cooool #dummysmiley: :-) :-P <3 and some
↪arrows < > -> <--"
tknzr.tokenize(s0)
```

```python
from nltk.stem import PorterStemmer

ps = PorterStemmer()

sample_words = ["marketing", "markets", "marketed", "marketer"]

print(sample_words)

for each in sample_words:
    print("{:s} -> {:s}".format(each, ps.stem(each)))
```

```python
from nltk.stem import WordNetLemmatizer
import nltk
nltk.download('wordnet')
wnl = WordNetLemmatizer()

print(wnl.lemmatize("beaten"))
print(wnl.lemmatize("beaten", "v"))
print(wnl.lemmatize("women", "n"))
print(wnl.lemmatize("happiest", "a"))
```

```
import nltk
nltk.download('averaged_perceptron_tagger')
nltk.download("tagsets")
from nltk.tokenize import TweetTokenizer
tknzr = TweetTokenizer()
s0 = "This is a cooool #dummysmiley: :-) :-P <3 and some
↪arrows < > -> <--"
tokens=tknzr.tokenize(s0)
tagged = nltk.pos_tag(tokens)
print(tagged)
```

```
import nltk.classify.util
from nltk.classify import NaiveBayesClassifier
from nltk.corpus import names


nltk.download("names")
def gender_features(word):
    return {'last_letter': word[-1]}


# Load data and training
names = ([(name, 'male') for name in names.words('male.txt')] +
         [(name, 'female') for name in names.words('female.txt
↪')])

#we will be using the last letter of each name as a feature
↪for training the model
featuresets = [(gender_features(n), g) for (n, g) in names]
train_set = featuresets
classifier = nltk.NaiveBayesClassifier.train(train_set)

print(names)
print("Anny")
print(classifier.classify(gender_features('Anny')))
```

Chapter 15
Image Processing

Abstract Image processing is a process of applying various operations on image data to extract useful information or produce an enhanced output. OpenCV, Scikit-image, and Pillow are useful tools that make image processing simpler. Techniques covered include generating histograms, extracting contours, grayscale transformation, histogram equalization, Fourier transformation, high pass filtering, and pattern recognition. A template matching example shown will be used to find and locate numbers and letters on a car plate.

Learning outcomes:

- Learn and apply basic image processing techniques.
- Write image processing pipeline with OpenCV in Python.
- Perform object detection through template matching.

This chapter provides an introduction to basic image processing techniques using the OpenCV computer vision library and some standard data analysis libraries in Python.

15.1 Load the Dependencies

This chapter requires the following libraries: **numpy**, **pandas**, **cv2**, **skimage**, **PIL**, **matplotlib**

```
import numpy as np
import pandas as pd
import cv2 as cv
#from google.colab.patches import cv2_imshow # for image
 ↪display
from PIL import Image
import matplotlib.pylab as plt
from skimage import data
from skimage.feature import match_template
from skimage.draw import circle
```

(continues on next page)

© The Author(s), under exclusive license to Springer Nature Singapore Pte Ltd. 2022
T. T. Teoh, Z. Rong, *Artificial Intelligence with Python*,
Machine Learning: Foundations, Methodologies, and Applications,
https://doi.org/10.1007/978-981-16-8615-3_15

(continued from previous page)

```
from skimage import io
from skimage import color
```

15.2   Load Image from urls

In this step we will read images from urls, and display them using openCV, please note the difference when reading image in RGB and BGR format. The default input color channels are in BGR format for openCV.

The following code allows us to show images in a Jupyter notebook and here is a brief walk through of what each step does:

- `io.imread`

  – read the picture as numerical array/matrixes

- `cv.cvtColor`

  – convert BGR into RGB
  – image when loaded by OpenCV is in BGR by default

- `cv.hconcat`

  – display images (BGR version and RGB version) and concatenate them horizontally

- `cv2_imshow` (for google colab). On local please use `matplotlib`

  – display images on our screen

```
# Create a list to store the urls of the images
urls = ["https://iiif.lib.ncsu.edu/iiif/0052574/full/800,/0/
↪default.jpg",
        "https://iiif.lib.ncsu.edu/iiif/0016007/full/800,/0/
↪default.jpg",
      "https://placekitten.com/800/571"]
# Read and display the image
# loop over the image URLs, you could store several image urls
↪in the list

for n, url in enumerate(urls):
  plt.figure()
  image = io.imread(url)
  image_2 = cv.cvtColor(image, cv.COLOR_BGR2RGB)
  final_frame = cv.hconcat((image, image_2))
  plt.imshow(final_frame)
  print('\n')
  plt.savefig(f'image_processing/img{n}.png')
```

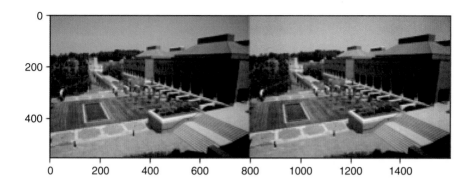

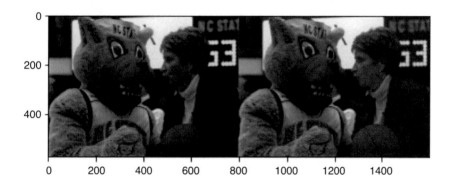

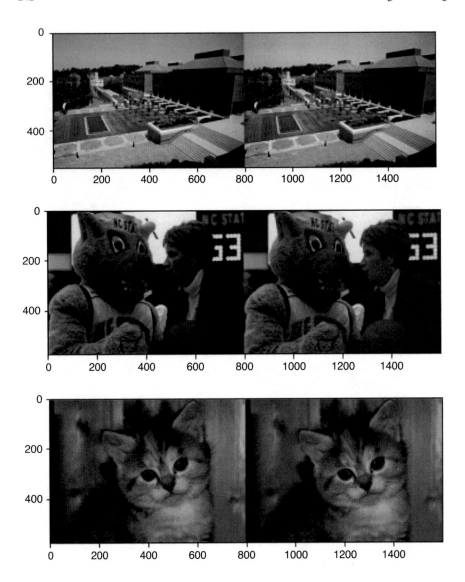

```
# Using Colab

# Create a list to store the urls of the images
urls = ["https://iiif.lib.ncsu.edu/iiif/0052574/full/800,/0/
↪default.jpg",
        "https://iiif.lib.ncsu.edu/iiif/0016007/full/800,/0/
↪default.jpg",
        "https://placekitten.com/800/571"]
# Read and display the image
```

(continues on next page)

(continued from previous page)

```
# loop over the image URLs, you could store several image urls
↪in the list

for n, url in enumerate(urls):
  plt.figure()
  image = io.imread(url)
  image_2 = cv.cvtColor(image, cv.COLOR_BGR2RGB)
  final_frame = cv.hconcat((image, image_2))
  # cv2_imshow(final_frame) // uncomment for colab
  print('\n')
plt.show()
```

```
<Figure size 432x288 with 0 Axes>
```

```
<Figure size 432x288 with 0 Axes>
```

```
<Figure size 432x288 with 0 Axes>
```

15.3   Image Analysis

Here we will analyze the image's contours and histograms. Firstly, let us take a look at some of the image's data.

Notice that a RGB image is 3 dimension in nature? Let us make sense of its shape and what the numbers represent.

```
# Check the image matrix data type (could know the bit depth
↪of the image)
io.imshow(image)
print(image.shape)
print(image.dtype)
# Check the height of image
print(image.shape[0])
# Check the width of image
print(image.shape[1])
# Check the number of channels of the image
print(image.shape[2])
plt.savefig(f'image_processing/img3.png')
```

```
(571, 800, 3)
uint8
571
800
3
```

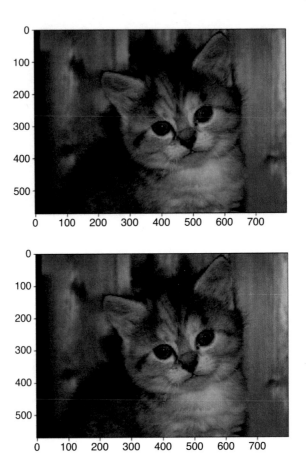

15.4   Image Histogram

Sometimes you want to enhance the contrast in your image or expand the contrast in a particular region while sacrificing the detail in colors that do not vary much, or do not matter. A good tool to find interesting regions is the histogram. To create a histogram of our image data, we use the matplot.pylab hist() function.

Display the histogram of all the pixels in the color image.

```
plt.hist(image.ravel(),bins = 256, range = [0,256])
plt.savefig(f'image_processing/img4.png')
```

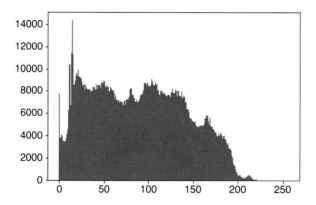

Display the histogram of R, G, B channel. We could observe that the green channel has many pixels in 255, which represents the white patch in the image.

```
color = ('b','g','r')
for i,col in enumerate(color):
    histr = cv.calcHist([image],[i],None,[256],[0,256])
    plt.plot(histr,color = col)
    plt.xlim([0,256])
plt.savefig(f'image_processing/img5.png')
```

```
gray_image = cv.cvtColor(image, cv.COLOR_BGR2GRAY)
plt.imshow(gray_image)
plt.savefig(f'image_processing/img6.png')
```

```
# Plot the histogram of the gray image. We could observe that␣
↪the frequency of
# the image hist has decreased ~ 1/3 of the histogram of color␣
↪image
```

(continues on next page)

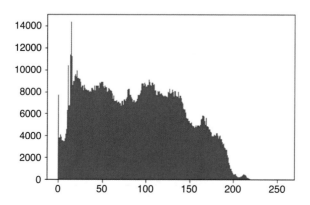

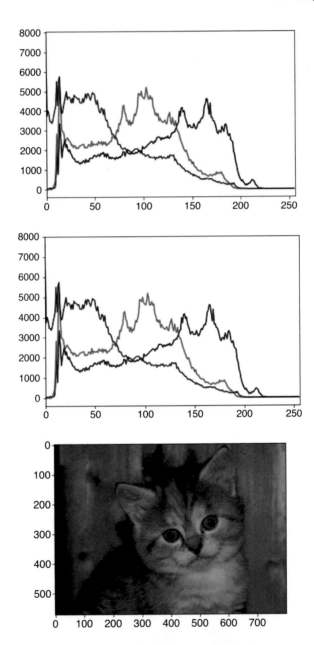

(continued from previous page)

```
plt.hist(gray_image.ravel(),bins = 256, range = [0, 256])
plt.savefig(f'image_processing/img7.png')
```

15.5   Contour

Contours can be explained simply as a curve joining all the continuous points (along the boundary), having same color or intensity. The contours are a useful tool for shape analysis and object detection and recognition.

Here is one method: Use the matplotlib contour. Refer to https://matplotlib.org/api/_as_gen/matplotlib.pyplot.contour.html for more details.

Notice that the edges of the cat is being highlighted here. matplotlib takes in the NumPy array and is able to return you the contours based on the origin.

```
plt.contour(gray_image, origin = "image")
plt.savefig(f'image_processing/img8.png')
```

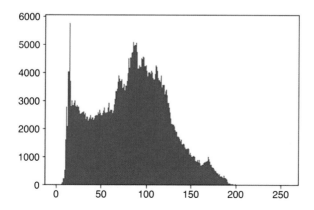

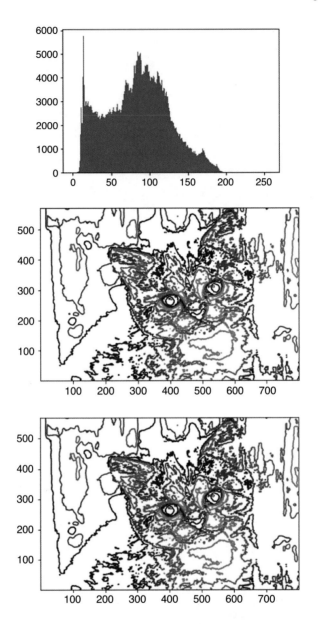

Another way would be to use opencv for contour finding. In OpenCV, finding contours is like finding white object from black background. So remember, object to be found should be white and background should be black.

See, there are three arguments in cv.findContours() function, first one is source image, second is contour retrieval mode, third is contour approximation method. And it outputs a modified image, contours, and hierarchy. Contours is a Python list of all the contours in the image. Each individual contour is a NumPy array of (x,y) coordinates of boundary points of the object.

```
ret, thresh = cv.threshold(gray_image,150,255,0)
contours, hierarchy = cv.findContours(thresh, cv.RETR_TREE, cv.
↪CHAIN_APPROX_SIMPLE)
image = cv.drawContours(image, contours, -1, (0, 255, 0), 3)

result = Image.fromarray((image).astype(np.uint8))
result.save('image_processing/img9.png')
```

15.6 Grayscale Transformation

The following section provides some examples of conducting mathematical transformations of the grayscale image.

This is an inverse operation of the grayscale image; you could see that the bright pixels become dark, and the dark pixels become bright.

```
im2 = - gray_image + 255
result = Image.fromarray((im2).astype(np.uint8))
result.save('image_processing/img10.png')
```

Another transform of the image, after adding a constant, all the pixels become brighter and a hazing-like effect of the image is generated.

• The lightness level of the gray_image decreases after this step.

```
im3 = gray_image + 50
result = Image.fromarray((im3).astype(np.uint8))
result.save('image_processing/img11.png')
```

15.7   Histogram Equalization

This section demonstrates histogram equalization on a dark image. This transform flattens the gray-level histogram so that all intensities are as equally common as possible. The transform function is a cumulative distribution function (cdf) of the pixel values in the image (normalized to map the range of pixel values to the desired range). This example uses image 4 (im4).

```
def histeq(im, nbr_bins = 256):
    """ Histogram equalization of a grayscale image.  """
    # get the image histogram
    imhist, bins = np.histogram(im.flatten(), nbr_bins, [0, 256])
    cdf = imhist.cumsum() # cumulative distribution function
    cdf = imhist.max()*cdf/cdf.max()   #normalize
    cdf_mask = np.ma.masked_equal(cdf, 0)
```

(continues on next page)

(continued from previous page)

```
  cdf_mask = (cdf_mask - cdf_mask.min())*255/(cdf_mask.max()-
↪cdf_mask.min())
  cdf = np.ma.filled(cdf_mask,0).astype('uint8')
  return cdf[im.astype('uint8')]

# apply the function on your dark image to increase the␣
↪contrast
# we could observe that the contrast of the black background␣
↪has increased
im5 = histeq(im3)
plt.imshow(im5)
plt.show()
```

15.8 Fourier Transformation

A Fourier transform is used to find the frequency domain of an image. You can consider an image as a signal which is sampled in two directions. So taking a Fourier transform in both X and Y directions gives you the frequency representation of image. For the sinusoidal signal, if the amplitude varies so fast in short time, you can say it is a high frequency signal. If it varies slowly, it is a low frequency signal. Edges and noises are high frequency contents in an image because they change drastically in images.

- Blur the grayscale image by a Gaussian filter with kernel size of 10

  – imBlur = cv.blur(gray_image,(5,5))

- Transform the image to frequency domain

  – f = np.fft.fft2(imBlur)

- Bring the zero-frequency component to the center

  – fshift = np.fft.fftshift(f)
  – magnitude_spectrum = 30*np.log(np.abs(fshift))

```
imBlur = cv.blur(gray_image,(5,5))
f = np.fft.fft2(imBlur)
fshift = np.fft.fftshift(f)
magnitude_spectrum = 30*np.log(np.abs(fshift))

plt.subplot(121),plt.imshow(imBlur, cmap = 'gray')
plt.title('Input Image'), plt.xticks([]), plt.yticks([])
plt.subplot(122),plt.imshow(magnitude_spectrum, cmap = 'gray')
plt.title('Magnitude Spectrum'), plt.xticks([]), plt.yticks([])
plt.show()
```

Input Image   Magnitude Spectrum

 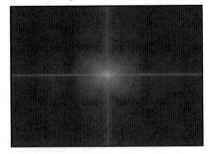

Input Image

Magnitude Spectrum

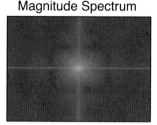

15.9 High pass Filtering in FFT

This section demonstrates conducting a high pass filter to remove the low frequency component, resulting in a sharpened image which contains the edges. Such technique allows us to find edges in the image.

```
rows, cols = imBlur.shape
crow,ccol = round(rows/2) , round(cols/2)
# remove low frequencies with a rectangle size of 10
fshift[crow-10:crow+10, ccol-10:ccol+10] = 0
f_ishift = np.fft.ifftshift(fshift)
img_back = np.fft.ifft2(f_ishift)
img_back = np.abs(img_back)

plt.figure(figsize=([20, 20]))
plt.subplot(131),plt.imshow(imBlur, cmap = 'gray')
plt.title('Input Image'), plt.xticks([]), plt.yticks([])
plt.subplot(132),plt.imshow(img_back, cmap = 'gray')
plt.title('Image after HPF'), plt.xticks([]), plt.yticks([])
plt.show()
```

Input Image

Image after HPF

Input Image   Image after HPF

15.10   Pattern Recognition

In the following section, we are going to go through pattern matching. Notice the below is a car plate number from this url: https://www.hdm-stuttgart.de/~maucher/Python/ComputerVision/html/_images/template.png. A copy of the image is available in dropbox as well. We would try to find individual alphabets. A common use case that is already widely implemented would be car plate number tracking.

```
from skimage import color
from skimage import io


full = color.rgb2gray(io.imread('./image_processing/platine.jpg
↪'))
plt.imshow(full,cmap = plt.cm.gray)
plt.title("Search pattern in this image")
```

```
ipykernel_launcher:5: FutureWarning: Non RGB image conversion␣
↪is now deprecated. For RGBA images, please use␣
↪rgb2gray(rgba2rgb(rgb)) instead. In version 0.19, a␣
↪ValueError will be raised if input image last dimension␣
↪length is not 3.
```

```
Text(0.5, 1.0, 'Search pattern in this image')
```

The following code crops out a portion of the image: `template = full[240:370,190:250]`. The numbers correspond to the width and height of the image. We are plotting the segment out. Here is what it looks like:

```
template = full[240:370,190:250]
plt.imshow(template,cmap = plt.cm.gray)
plt.figure()
plt.subplot(1,2,1)
plt.imshow(full,cmap = plt.cm.gray)
plt.title("Search pattern in this image")
```

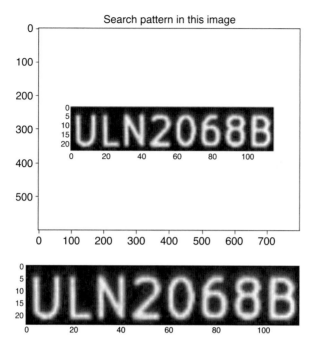

Text(0.5, 1.0, 'Search pattern in this image')

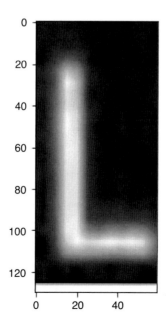

Search pattern in this image

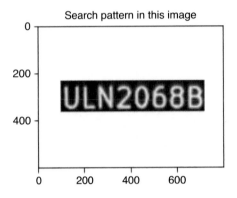

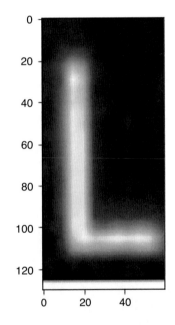

```
correlation=match_template(full,template)

xcoords=[]
ycoords=[]
for row in range(correlation.shape[0]):
        for col in range(correlation.shape[1]):
                if correlation[row,col]>0.9:
                        #print(row,col,correlation[row,col])
                        xcoords.append(col)
                        ycoords.append(row)
```

(continues on next page)

(continued from previous page)

```
plt.imshow(full,cmap = plt.cm.gray)
plt.title("Found patterns")
plt.plot(xcoords,ycoords,'om',ms=8,label="found matches")
plt.legend(loc=2,numpoints=1)
plt.legend()
plt.show()
```

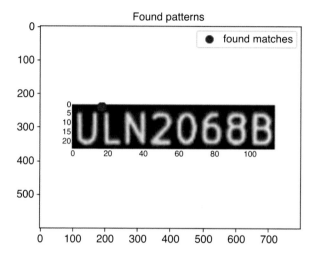

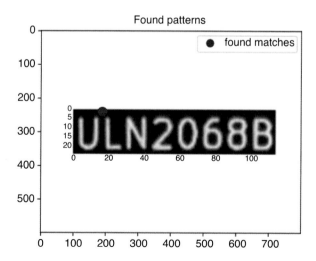

Notice that there is a mark at the top left hand corner of L in the image. This is because the L is being moved across the entire image and when there is a match it will be captured. You can try to change the shape and size of L, by rotating and resizing. The result might be different depending on the correlation between the template and the original image.

15.11   Sample Code

```
from skimage import io

image = io.imread("C:/Users/User/Dropbox/TT Library/AI Model/
↪Image & CNN/Pug.jpg")

io.imshow(image)
image.shape
```

```
import matplotlib.pylab as plt
plt.hist(image.flatten())
```

```
from skimage import color

imageGray = color.rgb2gray(image)
io.imshow(imageGray)
```

```
plt.contour(imageGray, origin = "image") #origin = "image",
↪else inverted
```

```
from skimage.transform import rescale

image_rescaled = rescale(imageGray, 0.25)
io.imshow(image_rescaled)
```

```
from skimage import filters
edges = filters.sobel(imageGray)
io.imshow(edges)
io.show()
```

```
import numpy as np

f = np.fft.fft2(imageGray)
# Bring the zero-frequency component to the center
fshift = np.fft.fftshift(f)
magnitude_spectrum = np.log(np.abs(fshift))

plt.imshow(imageGray, cmap = "gray")
```

```
plt.imshow(magnitude_spectrum)
plt.show()
```

```
import keras
from keras.datasets import mnist
from keras.models import Sequential
from keras.layers import Dense, Dropout, Flatten
from keras.layers import Conv2D, MaxPooling2D
import numpy as np

(x_train, y_train), (x_test, y_test) = mnist.load_data()

from skimage import io
io.imshow(x_train[0])

print(y_train[0])

print(x_train.shape)
print(x_train[0][0])
x_train = x_train.reshape(60000,28,28,1)
print("after", x_train[0][0])
x_test = x_test.reshape(10000,28,28,1)
```

```
print(y_train)
y_train = keras.utils.to_categorical(y_train, 10)
print("after", y_train[0])
y_test = keras.utils.to_categorical(y_test, 10)
```

```
model = Sequential()
model.add(Conv2D(32, (3,3), input_shape=(28,28,1)))
model.add(Conv2D(32, (3, 3)))
model.add(MaxPooling2D(pool_size=(2, 2)))
model.add(Dropout(0.25))
model.add(Flatten())
model.add(Dense(128))
model.add(Dense(10))
model.summary()
```

```
model.compile(loss=keras.losses.categorical_crossentropy,
↪metrics=['accuracy'])
```

```
model.fit(x_train, y_train, batch_size=5000, epochs=1)
score = model.evaluate(x_train, y_train)
print(score)
score = model.evaluate(x_test, y_test)
print(score)
```

Chapter 16
Convolutional Neural Networks

Abstract Convolutional Neural Networks are neural networks with convolution layers which perform operations similar to image processing filters. Convolutional Neural Networks are applied in a variety of tasks related to images such as image classification, object detection, and semantic segmentation. Popular Network architectures include ResNet, GoogleNet, and VGG. These networks are often trained on very large datasets, can be downloaded in Keras and Tensorflow, and can be later used for finetuning on other tasks.

Learning outcomes:

- Understand how convolution, pooling, and flattening operations are performed.
- Perform an image classification task using Convolutional Neural Networks.
- Familiarize with notable Convolution Neural Network Architectures.
- Understand Transfer Learning and Finetuning.
- Perform an image classification task through finetuning a Convolutional Neural Network previously trained on a separate task.
- Exposure to various applications of Convolutional Neural Networks.

A fully connected neural network consists of a series of fully connected layers that connect every neuron in one layer to every neuron in the other layer. The main problem with fully connected neural networks is that the number of weights required is very large for certain types of data. For example, an image of 224x224x3 would require 150528 weights in just the first hidden layer and will grow quickly for even bigger images. You can imagine how computationally intensive things would become once the images reach dimensions as large as 8K resolution images (7680×4320), training such a network would require a lot of time and resources.

However, for image data, repeating patterns can occur in different places. Hence we can train many smaller detectors, capable of sliding across an image, to take advantage of the repeating patterns. This would reduce the number of weights required to be trained.

A Convolutional Neural Network is a neural network with some convolutional layers (and some other layers). A convolutional layer has a number of filters that does the convolutional operation.

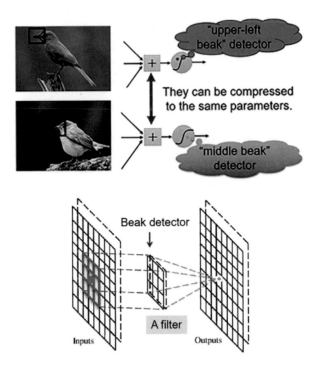

16.1 The Convolution Operation

The convolution operation is very similar to image processing filters such as the Sobel filter and Gaussian Filter. The Kernel slides across an image and multiplies the weights with each aligned pixel, element-wise across the filter. Afterwards the bias value is added to the output.

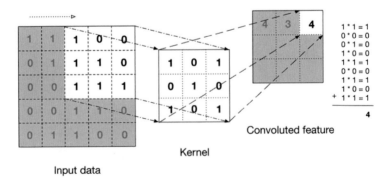

There are three hyperparameters deciding the spatial of the output feature map:

- Stride (S) is the step each time we slide the filter. When the stride is 1 then we move the filters one pixel at a time. When the stride is 2 (or uncommonly 3 or more, though this is rare in practice) then the filters jump 2 pixels at a time as we slide them around. This will produce smaller output volumes spatially.
- Padding (P): The inputs will be padded with a border of size according to the value specified. Most commonly, zero-padding is used to pad these locations. In neural network frameworks (caffe, TensorFlow, PyTorch, MXNet), the size of this zero-padding is a hyperparameter. The size of zero-padding can also be used to control the spatial size of the output volumes.
- Depth (D): The depth of the output volume is a hyperparameter too; it corresponds to the number of filters we use for a convolution layer.

Given w as the width of input, and F is the width of the filter, with P and S as padding, the output width will be: $(W+2P-F)/S+1$. Generally, set $P=(F-1)/2$ when the stride is S=1 ensures that the input volume and output volume will have the same size spatially.

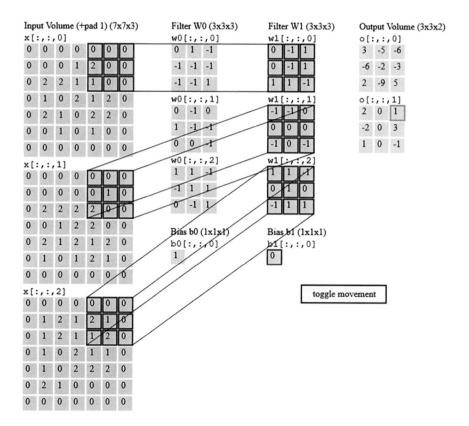

For an input of $7 \times 7 \times 3$ and an output depth of 2, we will have 6 kernels as shown below. Three for the first depth output and another 3 for the second depth output. The outputs of each filter are summed up to generate the output feature map.

In the example below, the output from each Kernel of Filter W1 is as follows:

Output of Kernel 1 = 1 Output of Kernel 2 = −2 Output of Kernel 3 = 2 Output of Filter W1 = Output of Kernel 1 + Output of Kernel 2 + Output of Kernel 3 + bias= $1 - 2 + 2 + 0 = 1$.

16.2 Pooling

Nowadays, a CNN always exploits extensive weight-sharing to reduce the degrees of the freedom of models. A pooling layer helps reduce computation time and gradually build up spatial and configural invariance. For image understanding, pooling layer helps extract more semantic meaning. The max pooling layer simply returns the maximum value over the values that the kernel operation is applied on. The example below illustrates the outputs of a max pooling and average pooling operation, respectively, given a kernel of size 2 and stride 2.

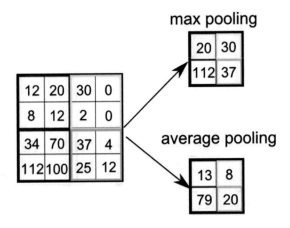

16.3 Flattening

Adding a Fully Connected layer is a (usually) cheap way of learning non-linear combinations of the high-level features as represented by the output of the convolutional layer. The Fully Connected layer is learning a possibly non-linear function in that space.

By flattening the image into a column vector, we have converted our input image into a suitable form for our Multi-Level Perceptron. The flattened output is fed to a feed-forward neural network and backpropagation applied to every iteration of training. Over a series of epochs, the model is able to distinguish between dominating and certain low-level features in images and classify them using the Softmax Classification technique.

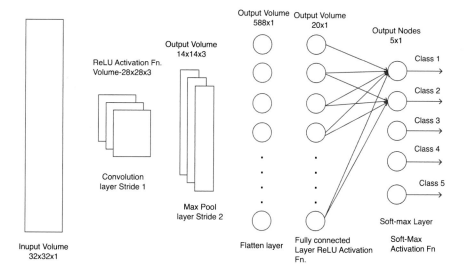

16.4 Exercise

We will build a small CNN using Convolution layers, Max Pooling layers, and Dropout layers in order to predict the type of fruit in a picture.

The dataset we will use is the fruits 360 dataset. You can obtain the dataset from this link: https://www.kaggle.com/moltean/fruits

```
import numpy as np # linear algebra
import pandas as pd # data processing, CSV file I/O (e.g. pd.
↪read_csv)
import os

from tensorflow.keras.models import Sequential
from tensorflow.keras.layers import Dense, Dropout, Flatten
from tensorflow.keras.layers import Conv2D, MaxPooling2D
from tensorflow.keras import optimizers
import numpy as np
import pandas as pd
```

(continues on next page)

(continued from previous page)

```
from tensorflow.keras.preprocessing.image import␣
↪ImageDataGenerator

import matplotlib.pyplot as plt
import matplotlib.image as mpimg
import pathlib
```

```
train_root  =pathlib.Path("D:/Programming Stuff/Teoh's Slides/
↪book-ai-potato (docs)/fruits-360_dataset/fruits-360/Training
↪")
test_root = pathlib.Path("D:/Programming Stuff/Teoh's Slides/
↪book-ai-potato (docs)/fruits-360_dataset/fruits-360/Test")
```

```
batch_size = 10
```

```
from skimage import io
image = io.imread("D:/Programming Stuff/Teoh's Slides//book-ai-
↪potato (docs)/fruits-360_dataset/fruits-360/Training/Apple␣
↪Braeburn/101_100.jpg")
print(image.shape)
print(image)
io.imshow(image)
```

```
(100, 100, 3)
[[[253 255 250]
  [255 255 251]
  [255 254 255]
  ...
  [255 255 255]
  [255 255 255]
  [255 255 255]]

 [[251 255 252]
  [253 255 252]
  [255 254 255]
  ...
  [255 255 255]
  [255 255 255]
  [255 255 255]]

 [[249 255 253]
  [251 255 254]
  [255 255 255]
  ...
  [255 255 255]
  [255 255 255]
  [255 255 255]]

 ...
```

(continues on next page)

(continued from previous page)

```
[[255 255 255]
 [255 255 255]
 [255 255 255]
 ...
 [255 255 255]
 [255 255 255]
 [255 255 255]]

[[255 255 255]
 [255 255 255]
 [255 255 255]
 ...
 [255 255 255]
 [255 255 255]
 [255 255 255]]

[[255 255 255]
 [255 255 255]
 [255 255 255]
 ...
 [255 255 255]
 [255 255 255]
 [255 255 255]]]
```

```
<matplotlib.image.AxesImage at 0x1543232f070>
```

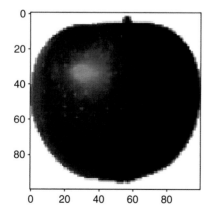

```
Generator = ImageDataGenerator()
train_data = Generator.flow_from_directory(train_root, (100,
↪100), batch_size=batch_size)
test_data = Generator.flow_from_directory(test_root, (100,
↪100), batch_size=batch_size)
```

```
Found 67692 images belonging to 131 classes.
Found 22688 images belonging to 131 classes.
```

```
num_classes = len([i for i in os.listdir(train_root)])
print(num_classes)
```

```
131
```

```
model = Sequential()

model.add(Conv2D(16, (5, 5), input_shape=(100, 100, 3),␣
↪activation='relu'))
model.add(MaxPooling2D(pool_size=(2, 2), strides=2))
model.add(Dropout(0.05))

model.add(Conv2D(32, (5, 5), activation='relu'))
model.add(MaxPooling2D(pool_size=(2, 2), strides=2))
model.add(Dropout(0.05))

model.add(Conv2D(64, (5, 5),activation='relu'))
model.add(MaxPooling2D(pool_size=(2, 2), strides=2))
model.add(Dropout(0.05))

model.add(Conv2D(128, (5, 5), activation='relu'))
model.add(MaxPooling2D(pool_size=(2, 2), strides=2))
model.add(Dropout(0.05))

model.add(Flatten())

model.add(Dense(1024, activation='relu'))
model.add(Dropout(0.05))

model.add(Dense(256, activation='relu'))
model.add(Dropout(0.05))

model.add(Dense(num_classes, activation="softmax"))
```

```
model.compile(loss=keras.losses.categorical_crossentropy,␣
↪optimizer=optimizers.Adam(), metrics=['accuracy'])
model.fit(train_data, batch_size = batch_size, epochs=2)
```

```
Epoch 1/2
6770/6770 [==============================] - 160s 24ms/step -␣
↪loss: 1.2582 - accuracy: 0.6622
Epoch 2/2
6770/6770 [==============================] - 129s 19ms/step -␣
↪loss: 0.5038 - accuracy: 0.8606
```

```
<tensorflow.python.keras.callbacks.History at 0x154323500a0>
```

```
score = model.evaluate(train_data)
print(score)
score = model.evaluate(test_data)
print(score)
```

```
6770/6770 [==============================] - 105s 15ms/step -⌣
↪loss: 0.2151 - accuracy: 0.9366
[0.21505890786647797, 0.9366099238395691]
2269/2269 [==============================] - 34s 15ms/step -⌣
↪loss: 0.8411 - accuracy: 0.8114
[0.8410834670066833, 0.8113980889320374]
```

16.5 CNN Architectures

There are various network architectures being used for image classification tasks. VGG16, Inception Net (GoogLeNet), and Resnet are some of the more notable ones.

16.5.1 VGG16

The VGG16 architecture garnered a lot of attention in 2014. It makes the improvement over its predecessor, AlexNet, through replacing large kernel-sized filters (11 and 5 in the first and second convolutional layer, respectively) with multiple 3×3 kernel-sized filters stacked together.

16.5.2 Inception Net

Before the Dense layers (which are placed at the end of the network), each time we add a new layer we face two main decisions:

1. Deciding whether we want to go with a Pooling or Convolutional operation;
2. Deciding the size and number of filters to be passed through the output of the previous layer.

Google researchers developed the Inception module allows us to apply different options all together in one single layer.

The main idea of the Inception module is that of running multiple operations (pooling, convolution) with multiple filter sizes (3x3, 5x5...) in parallel so that we do not have to face any trade-off.

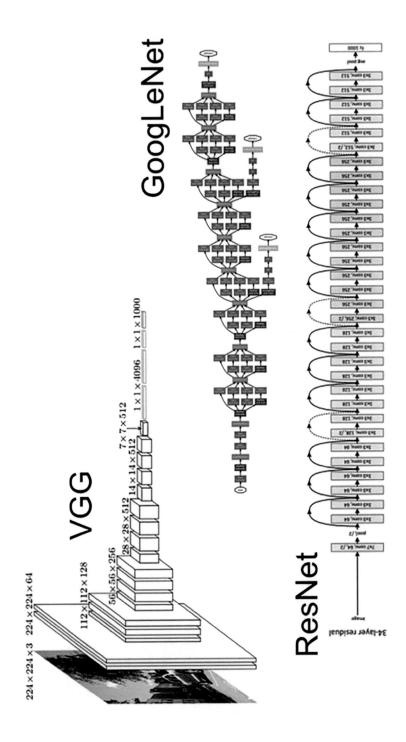

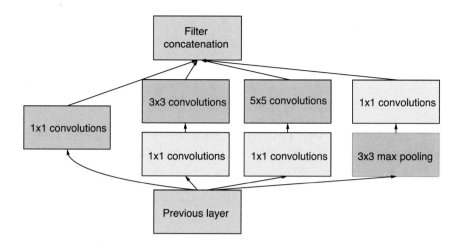

16.5.3 ResNet

Researchers thought that increasing more layers would improve the accuracy of the models. But there are two problems associated with it.

1. Vanishing gradient problem—Somewhat solved with regularization like batch normalization, etc. Gradients become increasingly smaller as the network becomes deeper, making it harder to train deep networks.
2. The authors observed that adding more layers did not improve the accuracy. Also, it is not over-fitting also as the training error is also increasing.

The basic intuition of the Residual connections is that, at each conv layer the network learns some features about the data F(x) and passes the remaining errors further into the network. So we can say the output error of the conv layer is H(x) = F(x) -x.

This solution also helped to alleviate the vanishing gradient problem as gradients can flow through the residual connections.

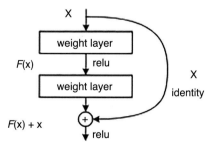

16.6   Finetuning

Neural networks are usually initialized with random weights. These weights will converge to some values after training for a series of epochs, to allow us to properly classify our input images. However, instead of a random initialization, we can initialize those weights to values that are already good to classify a different dataset.

Transfer Learning is the process of training a network that already performs well on one task, to perform a different task. Finetuning is an example of transfer learning, where we use another network trained on a much larger dataset to initialize and simply train it for classification. In finetuning, we can keep the weights of earlier layers as it has been observed that the Early layers contain more generic features, edges, color blobs and are more common to many visual tasks. Thus we can just Finetune the later layers which are more specific to the details of the class.

Through Transfer Learning, we would not require a dataset as big compared to having to train a network from scratch. We can reduce the required number of images from hundreds of thousands or even millions of images down to just a few thousands. Training Time is also sped up during the retraining process as it is much easier due to the initialization.

In the exercise below, we will finetune a ResNet50, pretrained on ImageNet (more than 14 million images, consisting of 1000 classes) for the same fruit classification task. In order to speed up the training process, we will freeze ResNet and simply train the last linear layer.

```
from tensorflow.keras.applications.resnet import ResNet50
resnet_model = ResNet50(include_top=False, weights='imagenet',␣
↪input_shape=(100,100,3))
resnet_model.trainable = False

from tensorflow.keras.layers import Conv2D, MaxPooling2D,␣
↪Flatten, Dense, Dropout, InputLayer, GlobalAveragePooling2D
from tensorflow.keras.models import Sequential
from tensorflow.keras import optimizers
model = Sequential()
model.add(resnet_model)
model.add(GlobalAveragePooling2D())
model.add(Dense(num_classes, activation='softmax'))
model.compile(loss=keras.losses.categorical_crossentropy,␣
↪optimizer=optimizers.Adam(), metrics=['accuracy'])
```

```
model.summary()
```

```
Model: "sequential_1"
_____
↪__
```

(continues on next page)

(continued from previous page)

```
Layer (type)                    Output Shape              Param #
=================================================================
resnet50 (Model)                (None, 4, 4, 2048)        23587712

↪__
global_average_pooling2d_1 (    (None, 2048)              0

↪__
dense_1 (Dense)                 (None, 131)               268419
=================================================================
Total params: 23,856,131
Trainable params: 268,419
Non-trainable params: 23,587,712

↪__
```

```
model.fit(train_data, epochs=1)
```

```
6770/6770 [==============================] - 388s 57ms/step -␣
↪loss: 0.1312 - accuracy: 0.9728
```

```
<tensorflow.python.keras.callbacks.History at 0x15420d63490>
```

```
score = model.evaluate(train_data)
print(score)
score = model.evaluate(test_data)
print(score)
```

```
6770/6770 [==============================] - 387s 57ms/step -␣
↪loss: 0.0214 - accuracy: 0.9927
[0.021364932879805565, 0.9926578998565674]
2269/2269 [==============================] - 132s 58ms/step -␣
↪loss: 0.3093 - accuracy: 0.9347
[0.3093399107456207, 0.9346791505813599]
```

16.7 Other Tasks That Use CNNs

CNNs are used in many other tasks apart from Image classification.

16.7.1   Object Detection

Classification tasks only tell us what is in the image and not where the object is. Object detection is the task of localizing objects within an image. CNNs, such as ResNets, are usually used as the feature extractor for object detection networks.

Groundtruth:
tv or monitor
tv or monitor (2)
tv or monitor (3)
person
remote control
remote control (2)

16.7.2   Semantic Segmentation

Using Fully Convolutional Nets, we can generate output maps which tell us which pixel belongs to which classes. This task is called Semantic Segmentation.

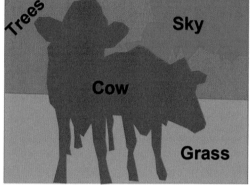

Chapter 17
Chatbot, Speech, and NLP

Abstract Chatbots are programs that are capable of conversing with people trained for their specific tasks, such as providing parents with information about a school. This chapter will provide the skills required to create a basic chatbot that can converse through speech. Speech to text tools will be used to convert speech data into text data. An encoder-decoder architecture model will be trained using Long-Short Term Memory units for a question and answer task for conversation.

Learning outcomes:

- Explore into speech to text capabilities in python.
- Represent text data in structured and easy-to-consume formats for chatbots.
- Familiarize with the Encoder-Decoder architecture.
- Develop a chatbot to answer questions.

In this chapter, we will explore the speech to text capabilities with python, then we will assemble a seq2seq LSTM model using Keras Functional API to create a working Chatbot which would answer questions asked to it. You can try integrating both programs together. However, do note that the code we have provided does not integrate both component.

Chatbots have become applications themselves. You can choose the field or stream and gather data regarding various questions. We can build a chatbot for an e-commerce website or a school website where parents could get information about the school.

Messaging platforms like Allo have implemented chatbot services to engage users. The famous Google Assistant, Siri, Cortana, and Alexa may have been build using similar models.

So, let us start building our Chatbot.

17.1   Speech to Text

```
#pip install SpeechRecognition
#pip install pipwin
#pipwin install pyaudio
import speech_recognition as sr
import sys
r = sr.Recognizer()

print("please say something in 4 seconds... and wait for 4
↪seconds for the answer.....")
print("Accessing Microphone..")

try:
    with sr.Microphone() as source:
        r.adjust_for_ambient_noise(source, duration=2)
    # use the default microphone as the audio source, duration
↪higher means environment noisier
        print("Waiting for you to speak...")
        audio = r.listen(source)                 # listen
↪for the first phrase and extract it into audio data

except (ModuleNotFoundError,AttributeError):
    print('Please check installation')
    sys.exit(0)

try:
    print("You said " + r.recognize_google(audio))     #
↪recognize speech using Google Speech Recognition
except LookupError:                              # speech is
↪unintelligible
    print("Could not understand audio")

except:
    print("Please retry...")
```

The following code requires `SpeechRecognition`, `pipwin`, and `pyaudio`. Please install first before carrying out the code. The code adjusts according to the ambient noise which help to capture what we have said. If there is an error, try adjusting the `duration` parameter in `adjust_for_ambient_noise`. Now we have managed to capture what is said in `r.recognize_google(audio)`. We will be able to use this can pass it to our chatbot.

17.2   Importing the Packages for Chatbot

We will import TensorFlow and our beloved Keras. Also, we import other modules which help in defining model layers.

```
import numpy as np
import tensorflow as tf
import pickle
from tensorflow.keras import layers , activations , models ,␣
↪preprocessing
```

17.3 Preprocessing the Data for Chatbot

17.3.1 Download the Data

The dataset hails from chatterbot/english on Kaggle.com by kausr25. It contains
pairs of questions and answers based on a number of subjects like food, history, AI,
etc.

The raw data could be found from this repo -> https://github.com/shubham0204/
Dataset_Archives

```
!wget https://github.com/shubham0204/Dataset_Archives/blob/
↪master/chatbot_nlp.zip?raw=true -O chatbot_nlp.zip
!unzip chatbot_nlp.zip
```

17.3.2 Reading the Data from the Files

We parse each of the .yaml files.

- Concatenate two or more sentences if the answer has two or more of them.
- Remove unwanted data types which are produced while parsing the data.
- Append <START> and <END> to all the answers.
- Create a Tokenizer and load the whole vocabulary (questions +
  answers) into it.

```
from tensorflow.keras import preprocessing , utils
import os
import yaml

dir_path = 'chatbot_nlp/data'
files_list = os.listdir(dir_path + os.sep)

questions = list()
answers = list()

for filepath in files_list:
    stream = open( dir_path + os.sep + filepath , 'rb')
```

(continues on next page)

(continued from previous page)

```
    docs = yaml.safe_load(stream)
    conversations = docs['conversations']
    for con in conversations:
        if len( con ) > 2 :
            questions.append(con[0])
            replies = con[ 1 : ]
            ans = ''
            for rep in replies:
                ans += ' ' + rep
            answers.append( ans )
        elif len( con )> 1:
            questions.append(con[0])
            answers.append(con[1])

answers_with_tags = list()
for i in range( len( answers ) ):
    if type( answers[i] ) == str:
        answers_with_tags.append( answers[i] )
    else:
        questions.pop( i )

answers = list()
for i in range( len( answers_with_tags ) ) :
    answers.append( '<START> ' + answers_with_tags[i] + ' <END>
↪' )

tokenizer = preprocessing.text.Tokenizer()
tokenizer.fit_on_texts( questions + answers )
VOCAB_SIZE = len( tokenizer.word_index )+1
print( 'VOCAB SIZE : {}'.format( VOCAB_SIZE ))
```

17.3.3 Preparing Data for Seq2Seq Model

Our model requires three arrays, namely encoder_input_data, decoder_input_data, and decoder_output_data.

For encoder_input_data:

- Tokenize the questions. Pad them to their maximum length.

For decoder_input_data:

- Tokenize the answers. Pad them to their maximum length.

For decoder_output_data:

- Tokenize the answers. Remove the first element from all the tokenized_answers. This is the <START> element which we added earlier.

```
from gensim.models import Word2Vec
import re

vocab = []
for word in tokenizer.word_index:
    vocab.append( word )

def tokenize( sentences ):
    tokens_list = []
    vocabulary = []
    for sentence in sentences:
        sentence = sentence.lower()
        sentence = re.sub( '[^a-zA-Z]', ' ', sentence )
        tokens = sentence.split()
        vocabulary += tokens
        tokens_list.append( tokens )
    return tokens_list , vocabulary

#p = tokenize( questions + answers )
#model = Word2Vec( p[ 0 ] )

#embedding_matrix = np.zeros( ( VOCAB_SIZE , 100 ) )
#for i in range( len( tokenizer.word_index ) ):
    #embedding_matrix[ i ] = model[ vocab[i] ]

# encoder_input_data
tokenized_questions = tokenizer.texts_to_sequences( questions )
maxlen_questions = max( [ len(x) for x in tokenized_questions
 ↪] )
padded_questions = preprocessing.sequence.pad_sequences( ↲
 ↪tokenized_questions , maxlen=maxlen_questions , padding='post
 ↪' )
encoder_input_data = np.array( padded_questions )
print( encoder_input_data.shape , maxlen_questions )

# decoder_input_data
tokenized_answers = tokenizer.texts_to_sequences( answers )
maxlen_answers = max( [ len(x) for x in tokenized_answers ] )
padded_answers = preprocessing.sequence.pad_sequences( ↲
 ↪tokenized_answers , maxlen=maxlen_answers , padding='post' )
decoder_input_data = np.array( padded_answers )
print( decoder_input_data.shape , maxlen_answers )

# decoder_output_data
tokenized_answers = tokenizer.texts_to_sequences( answers )
for i in range(len(tokenized_answers)) :
    tokenized_answers[i] = tokenized_answers[i][1:]
padded_answers = preprocessing.sequence.pad_sequences( ↲
 ↪tokenized_answers , maxlen=maxlen_answers , padding='post' )
onehot_answers = utils.to_categorical( padded_answers , VOCAB_
 ↪SIZE )
decoder_output_data = np.array( onehot_answers )
print( decoder_output_data.shape )
```

```
tokenized_questions[0],tokenized_questions[1]
```

```
padded_questions[0].shape
```

17.4   Defining the Encoder-Decoder Model

The model will have Embedding, LSTM, and Dense layers. The basic configuration is as follows.

- 2 Input Layers : One for `encoder_input_data` and another for `decoder_input_data`.
- Embedding layer : For converting token vectors to fix sized dense vectors. **(Note : Do not forget the `mask_zero=True` argument here)**
- LSTM layer : Provide access to Long-Short Term cells.

   Working :

1. The `encoder_input_data` comes in the Embedding layer (`encoder_embedding`).
2. The output of the Embedding layer goes to the LSTM cell which produces 2 state vectors (h and c which are `encoder_states`).
3. These states are set in the LSTM cell of the decoder.
4. The decoder_input_data comes in through the Embedding layer.
5. The Embeddings goes in LSTM cell (which had the states) to produce seqeunces.

   Image credits to Hackernoon.

```
encoder_inputs = tf.keras.layers.Input(shape=( maxlen_
↪questions , ))
encoder_embedding = tf.keras.layers.Embedding( VOCAB_SIZE, 200␣
↪, mask_zero=True ) (encoder_inputs)
encoder_outputs , state_h , state_c = tf.keras.layers.LSTM(␣
↪200 , return_state=True )( encoder_embedding )
encoder_states = [ state_h , state_c ]

decoder_inputs = tf.keras.layers.Input(shape=( maxlen_answers ,
↪ ))
decoder_embedding = tf.keras.layers.Embedding( VOCAB_SIZE, 200␣
↪, mask_zero=True) (decoder_inputs)
decoder_lstm = tf.keras.layers.LSTM( 200 , return_state=True ,␣
↪return_sequences=True )
decoder_outputs , _ , _ = decoder_lstm ( decoder_embedding ,␣
↪initial_state=encoder_states )
decoder_dense = tf.keras.layers.Dense( VOCAB_SIZE ,␣
↪activation=tf.keras.activations.softmax )
```

(continues on next page)

(continued from previous page)

```
output = decoder_dense ( decoder_outputs )

model = tf.keras.models.Model([encoder_inputs, decoder_inputs],
↪ output )
model.compile(optimizer=tf.keras.optimizers.RMSprop(), loss=
↪'categorical_crossentropy')

model.summary()
```

17.5   Training the Model

We train the model for a number of epochs with RMSprop optimizer and
categorical_crossentropy loss function.

```
model.fit([encoder_input_data , decoder_input_data], decoder_
↪output_data, batch_size=50, epochs=150, verbose=0 )
model.save( 'model.h5' )
```

```
output = model.predict([encoder_input_data[0,np.newaxis],␣
↪decoder_input_data[0,np.newaxis]])
```

```
output[0][0]
```

```
np.argmax(output[0][0])
```

```
tokenizer_dict = { tokenizer.word_index[i]:i for i in␣
↪tokenizer.word_index}
tokenizer_dict
```

```
tokenizer_dict[np.argmax(output[0][1])]
```

```
tokenizer_dict[np.argmax(output[0][2])]
```

```
output = model.predict([encoder_input_data[0,np.newaxis],␣
↪decoder_input_data[0,np.newaxis]])
sampled_word_indexes = np.argmax(output[0],1)
sentence = ""
maxlen_answers = 74
for sampled_word_index in sampled_word_indexes:
    sampled_word = None
    sampled_word = tokenizer_dict[sampled_word_index]
    sentence += ' {}'.format( sampled_word )
    if sampled_word == 'end' or len(sentence.split()) > maxlen_
↪answers:
```

(continues on next page)

(continued from previous page)

```
        break
sentence
```

```
def print_train_result(index):
    print(f"Question is : {questions[index]}")
    print(f"Answer is : {answers[index]}")
    output = model.predict([encoder_input_data[index,np.
↪newaxis], decoder_input_data[index,np.newaxis]])
    sampled_word_indexes = np.argmax(output[0],1)
    sentence = ""
    maxlen_answers = 74
    for sampled_word_index in sampled_word_indexes:
        sampled_word = None
        sampled_word = tokenizer_dict[sampled_word_index]
        sentence += ' {}'.format( sampled_word )
        if sampled_word == 'end' or len(sentence.split()) >⌴
↪maxlen_answers:
            break
    print(f"Model prediction: {sentence}")
```

```
print_train_result(4)
```

```
print_train_result(55)
```

```
print_train_result(32)
```

17.6 Defining Inference Models

We create inference models which help in predicting answers.

Encoder inference model: Takes the question as input and outputs LSTM states (h and c).

Decoder inference model: Takes in 2 inputs, one is the LSTM states (Output of encoder model), second is the answer input sequences (ones not having the <start> tag). It will output the answers for the question which we fed to the encoder model and its state values.

```
def make_inference_models():

    encoder_model = tf.keras.models.Model(encoder_inputs,⌴
↪encoder_states)

    decoder_state_input_h = tf.keras.layers.Input(shape=( 200 ,
↪))
    decoder_state_input_c = tf.keras.layers.Input(shape=( 200 ,
↪))
```

(continues on next page)

(continued from previous page)

```
    decoder_states_inputs = [decoder_state_input_h, decoder_
↪state_input_c]

    decoder_outputs, state_h, state_c = decoder_lstm(
        decoder_embedding , initial_state=decoder_states_
↪inputs)
    decoder_states = [state_h, state_c]
    decoder_outputs = decoder_dense(decoder_outputs)
    decoder_model = tf.keras.models.Model(
        [decoder_inputs] + decoder_states_inputs,
        [decoder_outputs] + decoder_states)

    return encoder_model , decoder_model
```

17.7 Talking with Our Chatbot

First, we define a method `str_to_tokens` which converts `str` questions to
Integer tokens with padding.

```
def str_to_tokens( sentence : str ):
    words = sentence.lower().split()
    tokens_list = list()
    for word in words:
        tokens_list.append( tokenizer.word_index[ word ] )
    return preprocessing.sequence.pad_sequences( [tokens_list]␣
↪, maxlen=maxlen_questions , padding='post')
```

1. First, we take a question as input and predict the state values using `enc_model`.

2. We set the state values in the decoder's LSTM.

3. Then, we generate a sequence which contains the `<start>` element.

4. We input this sequence in the `dec_model`.

5. We replace the `<start>` element with the element which was predicted by the
   `dec_model` and update the state values.

6. We carry out the above steps iteratively until we hit the `<end>` tag or the
   maximum answer length.

```
enc_model , dec_model = make_inference_models()

for _ in range(10):
    states_values = enc_model.predict( str_to_tokens( input(
↪'Enter question : ' ) ) )
    empty_target_seq = np.zeros( ( 1 , 1 ) )
    empty_target_seq[0, 0] = tokenizer.word_index['start']
```

(continues on next page)

(continued from previous page)

```
    stop_condition = False
    decoded_translation = ''
    while not stop_condition :
        dec_outputs , h , c = dec_model.predict ([ empty_target_
↪seq ] + states_values )
        sampled_word_index = np.argmax ( dec_outputs [0, -1, :] )
        sampled_word = None
        for word , index in tokenizer.word_index.items() :
            if sampled_word_index == index :
                decoded_translation += ' {}'.format ( word )
                sampled_word = word

        if sampled_word == 'end' or len (decoded_translation.
↪split()) > maxlen_answers:
            stop_condition = True

        empty_target_seq = np.zeros ( ( 1 , 1 ) )
        empty_target_seq[ 0 , 0 ] = sampled_word_index
        states_values = [ h , c ]

    print ( decoded_translation )
```

17.8   Sample Code

```
from IPython.display import Audio

#https://cloudconvert.com/m4a-to-wav
Path = "C:/Users/User/Dropbox/TT Library/AI Model/Speech &
↪Chatbot & NLP/Recording.wav"

Audio(Path)
```

```
import wave

audio = wave.open(Path)

from scipy.io import wavfile

fs, x = wavfile.read(Path)
print('Reading with scipy.io.wavfile.read:', x)
```

```
import speech_recognition as sr

r = sr.Recognizer()
audio1 = sr.AudioFile(Path)
with audio1 as source:
```

(continues on next page)

(continued from previous page)

```
    audio = r.record(source)

print("Speech to text : " + r.recognize_google(audio))
```

Chapter 18
Deep Convolutional Generative Adversarial Network

Abstract Deep convolutional generative adversarial networks consist of two models that are trained simultaneously by an adversarial process. A generator network learns to produce images that look real, while a discriminator network learns to tell real images apart from fakes. This process trains the generator network to generate real images that may not be found in the original dataset.

Learning outcomes:

- Understand the difference between generative and discriminative models.
- Understand the roles of the generator and discriminator in a GAN system.
- Train a GAN to generate new images.

18.1 What Are GANs?

Generative Adversarial Networks (GANs) are one of the most interesting ideas in computer science today. Two models are trained simultaneously by an adversarial process. A *generator* ("the artist") learns to create images that look real, while a *discriminator* ("the art critic") learns to tell real images apart from fakes.

During training, the *generator* progressively becomes better at creating images that look real, while the *discriminator* becomes better at telling them apart. The process reaches equilibrium when the *discriminator* can no longer distinguish real images from fakes.

This notebook demonstrates this process on the MNIST dataset. The following animation shows a series of images produced by the *generator* as it was trained for 50 epochs. The images begin as random noise and increasingly resemble handwritten digits over time.

To learn more about GANs, we recommend MIT's Intro to Deep Learning course.

© The Author(s), under exclusive license to Springer Nature Singapore Pte Ltd. 2022
T. T. Teoh, Z. Rong, *Artificial Intelligence with Python*,
Machine Learning: Foundations, Methodologies, and Applications,
https://doi.org/10.1007/978-981-16-8615-3_18

18.2   Setup

The following code represents the setup stage for dcgan.

```
import tensorflow as tf
```

```
tf.__version__
```

```
'2.3.0'
```

```
# To generate GIFs
!pip install -q imageio
!pip install -q git+https://github.com/tensorflow/docs
```

```
WARNING: You are using pip version 20.2.2; however, version 20.
↪2.3 is available.
You should consider upgrading via the '/tmpfs/src/tf_docs_env/
↪bin/python -m pip install --upgrade pip' command.
WARNING: You are using pip version 20.2.2; however, version 20.
↪2.3 is available.
You should consider upgrading via the '/tmpfs/src/tf_docs_env/
↪bin/python -m pip install --upgrade pip' command.
```

```
import glob
import imageio
import matplotlib.pyplot as plt
import numpy as np
import os
import PIL
from tensorflow.keras import layers
import time

from IPython import display
```

18.2.1   Load and Prepare the Dataset

You will use the MNIST dataset to train the generator and the discriminator. The generator will generate handwritten digits resembling the MNIST data.

```
(train_images, train_labels), (_, _) = tf.keras.datasets.mnist.
↪load_data()
```

```
train_images = train_images.reshape(train_images.shape[0], 28,␣
↪28, 1).astype('float32')
train_images = (train_images - 127.5) / 127.5 # Normalize the␣
↪images to [-1, 1]
```

```
BUFFER_SIZE = 60000
BATCH_SIZE = 256
```

```
# Batch and shuffle the data
train_dataset = tf.data.Dataset.from_tensor_slices(train_
↪images).shuffle(BUFFER_SIZE).batch(BATCH_SIZE)
```

18.3 Create the Models

Both the generator and discriminator are defined using the Keras Sequential API.

18.3.1 The Generator

The generator uses `tf.keras.layers.Conv2DTranspose` (upsampling) layers to produce an image from a seed (random noise). Start with a `Dense` layer that takes this seed as input, then upsample several times until you reach the desired image size of 28x28x1. Notice the `tf.keras.layers.LeakyReLU` activation for each layer, except the output layer, which uses tanh.

```
def make_generator_model():
    model = tf.keras.Sequential()
    model.add(layers.Dense(7*7*256, use_bias=False, input_
↪shape=(100,)))
    model.add(layers.BatchNormalization())
    model.add(layers.LeakyReLU())

    model.add(layers.Reshape((7, 7, 256)))
    assert model.output_shape == (None, 7, 7, 256) # Note:␣
↪None is the batch size

    model.add(layers.Conv2DTranspose(128, (5, 5), strides=(1,␣
↪1), padding='same', use_bias=False))
    assert model.output_shape == (None, 7, 7, 128)
    model.add(layers.BatchNormalization())
    model.add(layers.LeakyReLU())

    model.add(layers.Conv2DTranspose(64, (5, 5), strides=(2,␣
↪2), padding='same', use_bias=False))
```

(continues on next page)

(continued from previous page)

```
    assert model.output_shape == (None, 14, 14, 64)
    model.add(layers.BatchNormalization())
    model.add(layers.LeakyReLU())

    model.add(layers.Conv2DTranspose(1, (5, 5), strides=(2, 2),
↪ padding='same', use_bias=False, activation='tanh'))
    assert model.output_shape == (None, 28, 28, 1)

    return model
```

Use the (as yet untrained) generator to create an image.

```
generator = make_generator_model()

noise = tf.random.normal([1, 100])
generated_image = generator(noise, training=False)

plt.imshow(generated_image[0, :, :, 0], cmap='gray')
```

```
<matplotlib.image.AxesImage at 0x7f2729b9f6d8>
```

18.3.2 The Discriminator

The discriminator is a CNN-based image classifier.

```
def make_discriminator_model():
    model = tf.keras.Sequential()
    model.add(layers.Conv2D(64, (5, 5), strides=(2, 2),
↪padding='same',
                                        input_shape=[28, 28, 1]))
    model.add(layers.LeakyReLU())
    model.add(layers.Dropout(0.3))

    model.add(layers.Conv2D(128, (5, 5), strides=(2, 2),
↪padding='same'))
    model.add(layers.LeakyReLU())
    model.add(layers.Dropout(0.3))

    model.add(layers.Flatten())
    model.add(layers.Dense(1))

    return model
```

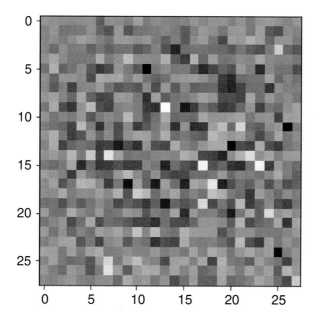

Use the (as yet untrained) discriminator to classify the generated images as real or fake. The model will be trained to output positive values for real images, and negative values for fake images.

```
discriminator = make_discriminator_model()
decision = discriminator(generated_image)
print (decision)
```

```
tf.Tensor([[0.0003284]], shape=(1, 1), dtype=float32)
```

18.4   Define the Loss and Optimizers

Define loss functions and optimizers for both models.

```
# This method returns a helper function to compute cross_
 ↪entropy loss
cross_entropy = tf.keras.losses.BinaryCrossentropy(from_
 ↪logits=True)
```

18.4.1 Discriminator Loss

This method quantifies how well the discriminator is able to distinguish real images from fakes. It compares the discriminator's predictions on real images to an array of 1s, and the discriminator's predictions on fake (generated) images to an array of 0s.

```
def discriminator_loss(real_output, fake_output):
    real_loss = cross_entropy(tf.ones_like(real_output), real_
↪output)
    fake_loss = cross_entropy(tf.zeros_like(fake_output), fake_
↪output)
    total_loss = real_loss + fake_loss
    return total_loss
```

18.4.2 Generator Loss

The generator's loss quantifies how well it was able to trick the discriminator. Intuitively, if the generator is performing well, the discriminator will classify the fake images as real (or 1). Here, we will compare the discriminator's decisions on the generated images to an array of 1s.

```
def generator_loss(fake_output):
    return cross_entropy(tf.ones_like(fake_output), fake_
↪output)
```

The discriminator and the generator optimizers are different since we will train two networks separately.

```
generator_optimizer = tf.keras.optimizers.Adam(1e-4)
discriminator_optimizer = tf.keras.optimizers.Adam(1e-4)
```

18.5 Save Checkpoints

This notebook also demonstrates how to save and restore models, which can be helpful in case a long running training task is interrupted.

```
checkpoint_dir = './training_checkpoints'
checkpoint_prefix = os.path.join(checkpoint_dir, "ckpt")
checkpoint = tf.train.Checkpoint(generator_optimizer=generator_
↪optimizer,
                                 discriminator_
↪optimizer=discriminator_optimizer,
                                 generator=generator,
                                 discriminator=discriminator)
```

18.6   Define the Training Loop

```
EPOCHS = 50
noise_dim = 100
num_examples_to_generate = 16

# We will reuse this seed overtime (so it's easier)
# to visualize progress in the animated GIF)
seed = tf.random.normal([num_examples_to_generate, noise_dim])
```

The training loop begins with generator receiving a random seed as input. That seed is used to produce an image. The discriminator is then used to classify real images (drawn from the training set) and fake images (produced by the generator). The loss is calculated for each of these models, and the gradients are used to update the generator and discriminator.

```
# Notice the use of `tf.function`
# This annotation causes the function to be "compiled".
@tf.function
def train_step(images):
    noise = tf.random.normal([BATCH_SIZE, noise_dim])

    with tf.GradientTape() as gen_tape, tf.GradientTape() as
↪disc_tape:
      generated_images = generator(noise, training=True)

      real_output = discriminator(images, training=True)
      fake_output = discriminator(generated_images,
↪training=True)

      gen_loss = generator_loss(fake_output)
      disc_loss = discriminator_loss(real_output, fake_output)

    gradients_of_generator = gen_tape.gradient(gen_loss,
↪generator.trainable_variables)
    gradients_of_discriminator = disc_tape.gradient(disc_loss,
↪discriminator.trainable_variables)

    generator_optimizer.apply_gradients(zip(gradients_of_
↪generator, generator.trainable_variables))
    discriminator_optimizer.apply_gradients(zip(gradients_of_
↪discriminator, discriminator.trainable_variables))
```

```
def train(dataset, epochs):
  for epoch in range(epochs):
    start = time.time()

    for image_batch in dataset:
      train_step(image_batch)
```

(continues on next page)

(continued from previous page)

```
    # Produce images for the GIF as we go
    display.clear_output(wait=True)
    generate_and_save_images(generator,
                             epoch + 1,
                             seed)

    # Save the model every 15 epochs
    if (epoch + 1) % 15 == 0:
      checkpoint.save(file_prefix = checkpoint_prefix)

    print ('Time for epoch {} is {} sec'.format(epoch + 1,␣
↪time.time()-start))

  # Generate after the final epoch
  display.clear_output(wait=True)
  generate_and_save_images(generator,
                           epochs,
                           seed)
```

Generate and save images.

```
def generate_and_save_images(model, epoch, test_input):
  # Notice `training` is set to False.
  # This is so all layers run in inference mode (batchnorm).
  predictions = model(test_input, training=False)

  fig = plt.figure(figsize=(4,4))

  for i in range(predictions.shape[0]):
      plt.subplot(4, 4, i+1)
      plt.imshow(predictions[i, :, :, 0] * 127.5 + 127.5, cmap=
↪'gray')
      plt.axis('off')

  plt.savefig('image_at_epoch_{:04d}.png'.format(epoch))
  plt.show()
```

18.6.1   Train the Model

Call the train() method defined above to train the generator and discriminator simultaneously. Note, training GANs can be tricky. It's important that the generator and discriminator do not overpower each other (e.g. that they train at a similar rate).

At the beginning of the training, the generated images look like random noise. As training progresses, the generated digits will look increasingly real. After about 50 epochs, they resemble MNIST digits. This may take about one minute/epoch with the default settings on Colab.

```
train(train_dataset, EPOCHS)
```

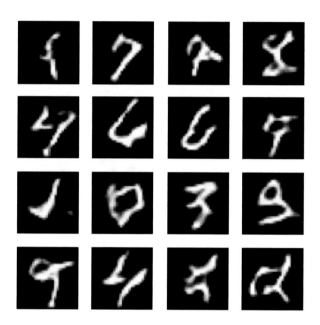

Restore the latest checkpoint.

```
checkpoint.restore(tf.train.latest_checkpoint(checkpoint_dir))
```

```
<tensorflow.python.training.tracking.util.CheckpointLoadStatus
↪at 0x7f2729bc3128>
```

18.6.2   Create a GIF

```
# Display a single image using the epoch number
def display_image(epoch_no):
  return PIL.Image.open('image_at_epoch_{:04d}.png'.
↪format(epoch_no))
```

```
display_image(EPOCHS)
```

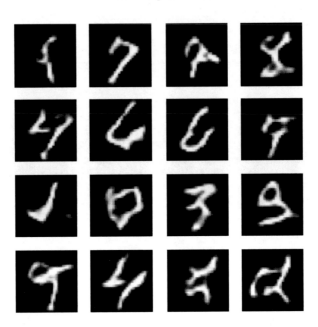

Use `imageio` to create an animated gif using the images saved during training.

```
anim_file = 'dcgan.gif'

with imageio.get_writer(anim_file, mode='I') as writer:
  filenames = glob.glob('image*.png')
  filenames = sorted(filenames)
  for filename in filenames:
    image = imageio.imread(filename)
    writer.append_data(image)
  image = imageio.imread(filename)
  writer.append_data(image)
```

```
import tensorflow_docs.vis.embed as embed
embed.embed_file(anim_file)
```

```
<IPython.core.display.HTML object>
```

Final Output
Epoch 0:

Epoch 10:

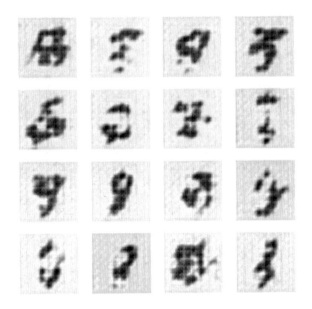

Epoch 30:

Epoch 50:

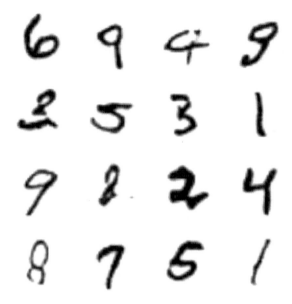

Notice after around 40 epochs the model learns how to generate digits.

Chapter 19
Neural Style Transfer

Abstract Neural style transfer takes in a content image and a style image to blend them together so that the output looks like the content image, but is painted in the style of the reference image. This can be done by using the features present in a previously trained network and well defined loss functions. A style loss is defined to represent how close the image is in terms of style to the reference. The content loss is defined to ensure important features of the original image is preserved.

Learning outcomes:

- Familiarize with Neural style transfer.
- Generate Style and Content representations.
- Perform style transfer.
- Reduce high frequency artifacts through regularization.

This tutorial uses deep learning to compose one image in the style of another image (ever wish you could paint like Picasso or Van Gogh?). This is known as *neural style transfer*, and the technique is outlined in A Neural Algorithm of Artistic Style (Gatys et al.).

Note: This tutorial demonstrates the original style-transfer algorithm. It optimizes the image content to a particular style. Modern approaches train a model to generate the stylized image directly (similar to cycleGAN). This approach is much faster (up to $1000\times$).

For a simple application of style transfer check out this tutorial to learn more about how to use the pretrained Arbitrary Image Stylization model from TensorFlow Hub or how to use a style-transfer model with TensorFlow Lite.

Neural style transfer is an optimization technique used to take two images—a *content* image and a *style reference* image (such as an artwork by a famous painter)—and blend them together so the output image looks like the content image, but "painted" in the style of the style reference image.

This is implemented by optimizing the output image to match the content statistics of the content image and the style statistics of the style reference image. These statistics are extracted from the images using a convolutional network.

T. T. Teoh, Z. Rong, *Artificial Intelligence with Python*,
Machine Learning: Foundations, Methodologies, and Applications,
https://doi.org/10.1007/978-981-16-8615-3_19

For example, let us take an image of this dog and Wassily Kandinsky's Composition 7:

Yellow Labrador Looking, from Wikimedia Commons by Elf. License CC BY-SA 3.0

Now how would it look like if Kandinsky decided to paint the picture of this Dog exclusively with this style? Something like this?

19.1   Setup

19.1.1   Import and Configure Modules

```
import os
import tensorflow as tf
# Load compressed models from tensorflow_hub
os.environ['TFHUB_MODEL_LOAD_FORMAT'] = 'COMPRESSED'
```

```
import IPython.display as display

import matplotlib.pyplot as plt
import matplotlib as mpl
mpl.rcParams['figure.figsize'] = (12,12)
mpl.rcParams['axes.grid'] = False

import numpy as np
import PIL.Image
import time
import functools
```

```
def tensor_to_image(tensor):
  tensor = tensor*255
  tensor = np.array(tensor, dtype=np.uint8)
  if np.ndim(tensor)>3:
    assert tensor.shape[0] == 1
    tensor = tensor[0]
  return PIL.Image.fromarray(tensor)
```

Download images and choose a style image and a content image:

```
content_path = tf.keras.utils.get_file('YellowLabradorLooking_
  ↪new.jpg', 'https://storage.googleapis.com/download.
  ↪tensorflow.org/example_images/YellowLabradorLooking_new.jpg')
style_path = tf.keras.utils.get_file('kandinsky5.jpg','https://
  ↪storage.googleapis.com/download.tensorflow.org/example_
  ↪images/Vassily_Kandinsky%2C_1913_-_Composition_7.jpg')
```

19.2   Visualize the Input

Define a function to load an image and limit its maximum dimension to 512 pixels.

```
def load_img(path_to_img):
  max_dim = 512
  img = tf.io.read_file(path_to_img)
  img = tf.image.decode_image(img, channels=3)
  img = tf.image.convert_image_dtype(img, tf.float32)

  shape = tf.cast(tf.shape(img)[:-1], tf.float32)
  long_dim = max(shape)
  scale = max_dim / long_dim

  new_shape = tf.cast(shape * scale, tf.int32)

  img = tf.image.resize(img, new_shape)
  img = img[tf.newaxis, :]
  return img
```

Create a simple function to display an image:

```
def imshow(image, title=None):
  if len(image.shape) > 3:
    image = tf.squeeze(image, axis=0)

  plt.imshow(image)
  if title:
    plt.title(title)
```

```
content_image = load_img(content_path)
style_image = load_img(style_path)

plt.subplot(1, 2, 1)
imshow(content_image, 'Content Image')

plt.subplot(1, 2, 2)
imshow(style_image, 'Style Image')
```

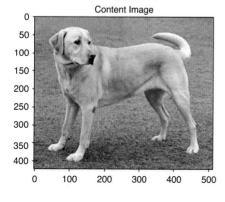
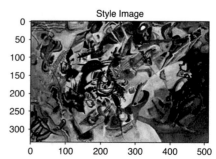

19.3   Fast Style Transfer Using TF-Hub

This tutorial demonstrates the original style-transfer algorithm, which optimizes the image content to a particular style. Before getting into the details, let us see how the TensorFlow Hub model does this:

```
import tensorflow_hub as hub
hub_model = hub.load('https://tfhub.dev/google/magenta/
↪arbitrary-image-stylization-v1-256/2')
stylized_image = hub_model(tf.constant(content_image), tf.
↪constant(style_image))[0]
tensor_to_image(stylized_image)
```

19.4   Define Content and Style Representations

Use the intermediate layers of the model to get the *content* and *style* representations of the image. Starting from the network's input layer, the first few layer activations represent low-level features like edges and textures. As you step through the network, the final few layers represent higher-level features—object parts like *wheels* or *eyes*. In this case, you are using the VGG19 network architecture, a pretrained image classification network. These intermediate layers are necessary to define the representation of content and style from the images. For an input image, try to match the corresponding style and content target representations at these intermediate layers.

Load a VGG19 and test run it on our image to ensure it is used correctly:

```
x = tf.keras.applications.vgg19.preprocess_input(content_
↪image*255)
x = tf.image.resize(x, (224, 224))
vgg = tf.keras.applications.VGG19(include_top=True, weights=
↪'imagenet')
prediction_probabilities = vgg(x)
prediction_probabilities.shape
```

```
TensorShape([1, 1000])
```

```
predicted_top_5 = tf.keras.applications.vgg19.decode_
↪predictions(prediction_probabilities.numpy())[0]
[(class_name, prob) for (number, class_name, prob) in_
↪predicted_top_5]
```

```
Downloading data from https://storage.googleapis.com/download.
↪tensorflow.org/data/imagenet_class_index.json
40960/35363 [==================================] - 0s 0us/step
```

```
[('Labrador_retriever', 0.49317262),
 ('golden_retriever', 0.23665187),
 ('kuvasz', 0.036357313),
 ('Chesapeake_Bay_retriever', 0.024182774),
 ('Greater_Swiss_Mountain_dog', 0.018646035)]
```

Now load a VGG19 without the classification head, and list the layer names

```
vgg = tf.keras.applications.VGG19(include_top=False, weights=
↪'imagenet')

print()
for layer in vgg.layers:
  print(layer.name)
```

```
Downloading data from https://storage.googleapis.com/
↪tensorflow/keras-applications/vgg19/vgg19_weights_tf_dim_
↪ordering_tf_kernels_notop.h5
80142336/80134624 [==============================] - 2s 0us/
↪step

input_2
block1_conv1
block1_conv2
block1_pool
block2_conv1
block2_conv2
block2_pool
block3_conv1
```

(continues on next page)

(continued from previous page)

```
block3_conv2
block3_conv3
block3_conv4
block3_pool
block4_conv1
block4_conv2
block4_conv3
block4_conv4
block4_pool
block5_conv1
block5_conv2
block5_conv3
block5_conv4
block5_pool
```

Choose intermediate layers from the network to represent the style and content of the image:

```
content_layers = ['block5_conv2']

style_layers = ['block1_conv1',
                'block2_conv1',
                'block3_conv1',
                'block4_conv1',
                'block5_conv1']

num_content_layers = len(content_layers)
num_style_layers = len(style_layers)
```

19.4.1 Intermediate Layers for Style and Content

So why do these intermediate outputs within our pretrained image classification network allow us to define style and content representations?

At a high level, in order for a network to perform image classification (which this network has been trained to do), it must understand the image. This requires taking the raw image as input pixels and building an internal representation that converts the raw image pixels into a complex understanding of the features present within the image.

This is also a reason why convolutional neural networks are able to generalize well: they are able to capture the invariances and defining features within classes (e.g. cats vs. dogs) that are agnostic to background noise and other nuisances. Thus, somewhere between where the raw image is fed into the model and the output classification label, the model serves as a complex feature extractor. By accessing intermediate layers of the model, you are able to describe the content and style of input images.

19.5 Build the Model

The networks in `tf.keras.applications` are designed so you can easily extract the intermediate layer values using the Keras functional API.

To define a model using the functional API, specify the inputs and outputs:

`model = Model(inputs, outputs)`

This following function builds a VGG19 model that returns a list of intermediate layer outputs:

```
def vgg_layers(layer_names):
  """ Creates a vgg model that returns a list of intermediate
↪output values."""
  # Load our model. Load pretrained VGG, trained on imagenet
↪data
  vgg = tf.keras.applications.VGG19(include_top=False, weights=
↪'imagenet')
  vgg.trainable = False

  outputs = [vgg.get_layer(name).output for name in layer_
↪names]

  model = tf.keras.Model([vgg.input], outputs)
  return model
```

And to create the model:

```
style_extractor = vgg_layers(style_layers)
style_outputs = style_extractor(style_image*255)

#Look at the statistics of each layer's output
for name, output in zip(style_layers, style_outputs):
  print(name)
  print("   shape: ", output.numpy().shape)
  print("   min: ", output.numpy().min())
  print("   max: ", output.numpy().max())
  print("   mean: ", output.numpy().mean())
  print()
```

```
block1_conv1
  shape:  (1, 336, 512, 64)
  min:   0.0
  max:   835.5255
  mean:  33.97525

block2_conv1
  shape:  (1, 168, 256, 128)
  min:   0.0
  max:   4625.8867
  mean:  199.82687
```

(continues on next page)

(continued from previous page)

```
block3_conv1
  shape:   (1, 84, 128, 256)
  min:   0.0
  max:   8789.24
  mean:   230.78099

block4_conv1
  shape:   (1, 42, 64, 512)
  min:   0.0
  max:   21566.133
  mean:   791.24005

block5_conv1
  shape:   (1, 21, 32, 512)
  min:   0.0
  max:   3189.2532
  mean:   59.179478
```

19.6   Calculate Style

The content of an image is represented by the values of the intermediate feature maps.

It turns out, the style of an image can be described by the means and correlations across the different feature maps. Calculate a Gram matrix that includes this information by taking the outer product of the feature vector with itself at each location, and averaging that outer product over all locations. This Gram matrix can be calculated for a particular layer as

$$G_{cd}^l = \frac{\sum_{ij} F_{ijc}^l(x) F_{ijd}^l(x)}{IJ}$$

This can be implemented concisely using the $\tt{tf.linalg.einsum}$ function:

```
def gram_matrix(input_tensor):
  result = tf.linalg.einsum('bijc,bijd->bcd', input_tensor,␣
  ↪input_tensor)
  input_shape = tf.shape(input_tensor)
  num_locations = tf.cast(input_shape[1]*input_shape[2], tf.
  ↪float32)
  return result/(num_locations)
```

19.7   Extract Style and Content

Build a model that returns the style and content tensors.

```python
class StyleContentModel(tf.keras.models.Model):
  def __init__(self, style_layers, content_layers):
    super(StyleContentModel, self).__init__()
    self.vgg =  vgg_layers(style_layers + content_layers)
    self.style_layers = style_layers
    self.content_layers = content_layers
    self.num_style_layers = len(style_layers)
    self.vgg.trainable = False

  def call(self, inputs):
    "Expects float input in [0,1]"
    inputs = inputs*255.0
    preprocessed_input = tf.keras.applications.vgg19.
↪preprocess_input(inputs)
    outputs = self.vgg(preprocessed_input)
    style_outputs, content_outputs = (outputs[:self.num_style_
↪layers],
                                      outputs[self.num_style_
↪layers:])

    style_outputs = [gram_matrix(style_output)
                     for style_output in style_outputs]

    content_dict = {content_name:value
                    for content_name, value
                    in zip(self.content_layers, content_
↪outputs)}

    style_dict = {style_name:value
                  for style_name, value
                  in zip(self.style_layers, style_outputs)}

    return {'content':content_dict, 'style':style_dict}
```

When called on an image, this model returns the gram matrix (style) of the
`style_layers` and content of the `content_layers`:

```python
extractor = StyleContentModel(style_layers, content_layers)

results = extractor(tf.constant(content_image))

print('Styles:')
for name, output in sorted(results['style'].items()):
  print("  ", name)
  print("    shape: ", output.numpy().shape)
  print("    min: ", output.numpy().min())
  print("    max: ", output.numpy().max())
  print("    mean: ", output.numpy().mean())
```

(continues on next page)

(continued from previous page)

```
  print()

print("Contents:")
for name, output in sorted(results['content'].items()):
  print("  ", name)
  print("      shape: ", output.numpy().shape)
  print("      min: ", output.numpy().min())
  print("      max: ", output.numpy().max())
  print("      mean: ", output.numpy().mean())
```

```
Styles:
  block1_conv1
    shape: (1, 64, 64)
    min: 0.005522847
    max: 28014.559
    mean: 263.79025

  block2_conv1
    shape: (1, 128, 128)
    min: 0.0
    max: 61479.49
    mean: 9100.949

  block3_conv1
    shape: (1, 256, 256)
    min: 0.0
    max: 545623.44
    mean: 7660.9766

  block4_conv1
    shape: (1, 512, 512)
    min: 0.0
    max: 4320501.0
    mean: 134288.86

  block5_conv1
    shape: (1, 512, 512)
    min: 0.0
    max: 110005.38
    mean: 1487.0381

Contents:
  block5_conv2
    shape: (1, 26, 32, 512)
    min: 0.0
    max: 2410.8796
    mean: 13.764152
```

19.8 Run Gradient Descent

With this style and content extractor, you can now implement the style-transfer
algorithm. Do this by calculating the mean square error for your image's output
relative to each target, then take the weighted sum of these losses.

Set your style and content target values:

```
style_targets = extractor(style_image)['style']
content_targets = extractor(content_image)['content']
```

Define a tf.Variable to contain the image to optimize. To make this quick,
initialize it with the content image (the tf.Variable must be the same shape as
the content image):

```
image = tf.Variable(content_image)
```

Since this is a float image, define a function to keep the pixel values between 0
and 1:

```
def clip_0_1(image):
  return tf.clip_by_value(image, clip_value_min=0.0, clip_
↪value_max=1.0)
```

Create an optimizer. The paper recommends LBFGS, but Adam works okay, too:

```
opt = tf.optimizers.Adam(learning_rate=0.02, beta_1=0.99,␣
↪epsilon=1e-1)
```

To optimize this, use a weighted combination of the two losses to get the total
loss:

```
style_weight=1e-2
content_weight=1e4
```

```
def style_content_loss(outputs):
    style_outputs = outputs['style']
    content_outputs = outputs['content']
    style_loss = tf.add_n([tf.reduce_mean((style_outputs[name] -
↪style_targets[name])**2)
                               for name in style_outputs.keys()])
    style_loss *= style_weight / num_style_layers

    content_loss = tf.add_n([tf.reduce_mean((content_
↪outputs[name]-content_targets[name])**2)
                               for name in content_outputs.
↪keys()])
    content_loss *= content_weight / num_content_layers
    loss = style_loss + content_loss
    return loss
```

Use tf.GradientTape to update the image.

```
@tf.function()
def train_step(image):
  with tf.GradientTape() as tape:
    outputs = extractor(image)
    loss = style_content_loss(outputs)

  grad = tape.gradient(loss, image)
  opt.apply_gradients([(grad, image)])
  image.assign(clip_0_1(image))
```

Now run a few steps to test:

```
train_step(image)
train_step(image)
train_step(image)
tensor_to_image(image)
```

Since it is working, perform a longer optimization:

```
import time
start = time.time()

epochs = 10
steps_per_epoch = 100

step = 0
for n in range(epochs):
  for m in range(steps_per_epoch):
    step += 1
```

(continues on next page)

(continued from previous page)

```
      train_step(image)
      print(".", end='')
  display.clear_output(wait=True)
  display.display(tensor_to_image(image))
  print("Train step: {}".format(step))

end = time.time()
print("Total time: {:.1f}".format(end-start))
```

```
Train step: 1000
Total time: 20.3
```

19.9 Total Variation Loss

One downside to this basic implementation is that it produces a lot of high frequency artifacts. Decrease these using an explicit regularization term on the high frequency components of the image. In style transfer, this is often called the *total variation loss*:

```
def high_pass_x_y(image):
  x_var = image[:,:,1:,:] - image[:,:,:-1,:]
  y_var = image[:,1:,:,:] - image[:,:-1,:,:]

  return x_var, y_var
```

```
x_deltas, y_deltas = high_pass_x_y(content_image)

plt.figure(figsize=(14,10))
plt.subplot(2,2,1)
imshow(clip_0_1(2*y_deltas+0.5), "Horizontal Deltas: Original")

plt.subplot(2,2,2)
imshow(clip_0_1(2*x_deltas+0.5), "Vertical Deltas: Original")

x_deltas, y_deltas = high_pass_x_y(image)

plt.subplot(2,2,3)
imshow(clip_0_1(2*y_deltas+0.5), "Horizontal Deltas: Styled")

plt.subplot(2,2,4)
imshow(clip_0_1(2*x_deltas+0.5), "Vertical Deltas: Styled")
```

This shows how the high frequency components have increased.

Also, this high frequency component is basically an edge detector. You can get similar output from the Sobel edge detector, for example:

```
plt.figure(figsize=(14,10))

sobel = tf.image.sobel_edges(content_image)
plt.subplot(1,2,1)
imshow(clip_0_1(sobel[...,0]/4+0.5), "Horizontal Sobel-edges")
plt.subplot(1,2,2)
imshow(clip_0_1(sobel[...,1]/4+0.5), "Vertical Sobel-edges")
```

The regularization loss associated with this is the sum of the squares of the values:

```
def total_variation_loss(image):
  x_deltas, y_deltas = high_pass_x_y(image)
  return tf.reduce_sum(tf.abs(x_deltas)) + tf.reduce_sum(tf.
  ↪abs(y_deltas))
```

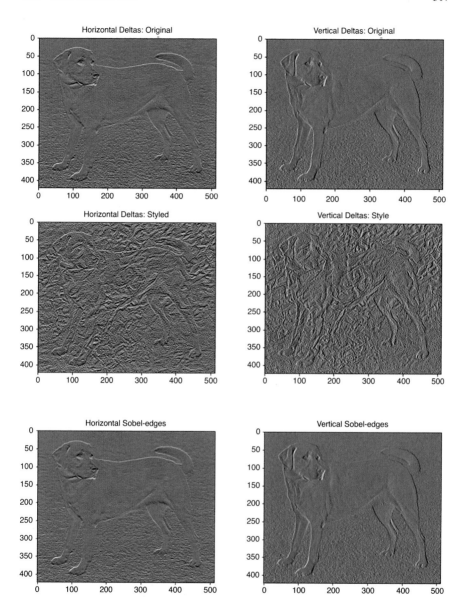

```
total_variation_loss(image).numpy()
```

```
149362.55
```

That demonstrated what it does. But there is no need to implement it yourself, TensorFlow includes a standard implementation:

```
tf.image.total_variation(image).numpy()
```

```
array([149362.55], dtype=float32)
```

19.10   Re-run the Optimization

Choose a weight for the `total_variation_loss`:

```
total_variation_weight=30
```

Now include it in the `train_step` function:

```
@tf.function()
def train_step(image):
  with tf.GradientTape() as tape:
    outputs = extractor(image)
    loss = style_content_loss(outputs)
    loss += total_variation_weight*tf.image.total_
↪variation(image)

    grad = tape.gradient(loss, image)
    opt.apply_gradients([(grad, image)])
    image.assign(clip_0_1(image))
```

Reinitialize the optimization variable:

```
image = tf.Variable(content_image)
```

And run the optimization:

```
import time
start = time.time()

epochs = 10
steps_per_epoch = 100

step = 0
for n in range(epochs):
  for m in range(steps_per_epoch):
    step += 1
    train_step(image)
    print(".", end='')
  display.clear_output(wait=True)
  display.display(tensor_to_image(image))
  print("Train step: {}".format(step))

end = time.time()
print("Total time: {:.1f}".format(end-start))
```

```
Train step: 1000
Total time: 21.4
```

Finally, save the result:

```python
file_name = 'stylized-image.png'
tensor_to_image(image).save(file_name)

try:
  from google.colab import files
except ImportError:
    pass
else:
  files.download(file_name)
```

Chapter 20
Reinforcement Learning

Abstract Reinforcement learning consists of developing an algorithm capable of achieving a particular objective through reward and punishment. A state is defined to allow the algorithm to understand its current circumstance and choose specific actions that it can take. Q-learning is the process of updating a Q-table to record the maximum expected future rewards. The algorithm will have to balance between exploration and exploitation in the learning process.

Learning outcomes:

- Introduction to the theories of reinforcement learning.
- Exploration of how Q-learning works.
- What is the next step after Q-learning.

Most of us would probably have heard of AI learning to play computer games on its own. A very popular example would be Deepmind. Deepmind took the world by surprise when its AlphaGo program won the Go world champion. In recent times, AI have been able to defeat human players in strategy game. One such example would be OpenAI's AlphaStar. Here, the difficulty is compounded as such game requires long term strategic planning.

Dario "TLO" Wünsch, a professional StarCraft player, remarked "I've found AlphaStar's gameplay incredibly impressive—the system is very skilled at assessing its strategic position, and knows exactly when to engage or disengage with its opponent. And while AlphaStar has excellent and precise control, it doesn't feel superhuman—certainly not on a level that a human couldn't theoretically achieve. Overall, it feels very fair—like it is playing a 'real' game of StarCraft."

20.1 Reinforcement Learning Analogy

Consider the scenario of teaching a dog new tricks. The dog doesn't understand human language, so we can't tell him what to do. Instead, we can create a situation or a cue, and the dog tries to behave in different ways. If the dog's response is desired, we reward them with their favorite snack. Now guess what, the next time

T. T. Teoh, Z. Rong, *Artificial Intelligence with Python*,
Machine Learning: Foundations, Methodologies, and Applications,
https://doi.org/10.1007/978-981-16-8615-3_20

the dog is exposed to the same situation, the dog executes a similar action with even more enthusiasm in expectation of more food. That's like learning "what to do" from positive experiences. Similarly, dogs will tend to learn what not to do when face with negative experiences. For example, whenever the dog behaves undesirably, we would admonish it. This helps the dog to understand and reinforce behavior that is desirable. At the same time, the dog would avoid undesirable behavior.

That's exactly how Reinforcement Learning works in a broader sense:

- Your dog is an "agent" that is exposed to the environment. The environment could in your house, with you.
- The situations they encounter are analogous to a state. An example of a state could be your dog standing and you use a specific word in a certain tone in your living room.
- Our agents react by performing an action to transition from one "state" to another "state," your dog goes from standing to sitting, for example. After the transition, they may receive a reward or penalty in return. You give them a treat! Or a "No" as a penalty. The policy is the strategy of choosing an action given a state in expectation of better outcomes.

Here are some points to take note of:

- Greedy (pursuit of current rewards) is not always good.

  - There are things that are easy to do for instant gratification, and there's things that provide long term rewards. The goal is to not be greedy by looking for the quick immediate rewards, but instead to optimize for maximum rewards over the whole training.

- Sequence matters in Reinforcement Learning

  - The reward agent does not just depend on the current state but the entire history of states. Unlike supervised, timestep and sequence of state–action–reward is important here.

20.2 Q-learning

In our example below, we will be using OpenAI Gym's Taxi environment.

```
import sys
sys.tracebacklimit = 0
import gym
import numpy as np
import random
from IPython.display import clear_output
from IPython.display import Markdown, display
def printmd(string):
    display(Markdown(string))
```

(continues on next page)

(continued from previous page)

```
# Init Taxi-V2 Env
env = gym.make("Taxi-v3").env

# Init arbitrary values
q_table = np.zeros([env.observation_space.n, env.action_space.
↪n])

# Hyperparameters
alpha = 0.7 # Momentum 0.2, Current 0.8 Greedy, 0.2 is to␣
↪reduce volatility and flip flop
gamma = 0.2 # Learning Rate 0.1 Greediness is 10%
epsilon = 0.4 # explore 10% exploit 90%


all_epochs = []
all_penalties = []
training_memory = []

for i in range(1, 50000):
    state = env.reset()

    # Init Vars
    epochs, penalties, reward, = 0, 0, 0
    done = False

    #training
    while not done:
        if random.uniform(0, 1) < epsilon:
            # Check the action space
            action = env.action_space.sample() # for explore
        else:
            # Check the learned values
            action = np.argmax(q_table[state]) # for exploit

        next_state, reward, done, info = env.step(action) #gym␣
↪generate, the environment already setup for you

        old_value = q_table[state, action]
        next_max = np.max(q_table[next_state]) #take highest␣
↪from q table for exploit

        # Update the new value
        new_value = (1 - alpha) * old_value + alpha * \
            (reward + gamma * next_max)
        q_table[state, action] = new_value

        # penalty for performance evaluation
        if reward == -10:
            penalties += 1
```

(continues on next page)

(continued from previous page)

```
            state = next_state
            epochs += 1


     if i % 100 == 0:
            training_memory.append(q_table.copy())
            clear_output(wait=True)
            print("Episode:", i)
            print("Saved q_table during training:", i)

print("Training finished.")
print(q_table)
```

```
Episode: 49900
Saved q_table during training: 49900
Training finished.
[[  0.            0.            0.            0.            0.
     0.         ]
 [ -1.24999956  -1.24999782  -1.24999956  -1.24999782  -1.
 ↪24998912
   -10.24999782]
 [ -1.249728    -1.24864     -1.249728    -1.24864     -1.2432
   -10.24864    ]
 ...
 [ -1.2432      -1.216       -1.2432      -1.24864    -10.2432
   -10.2432     ]
 [ -1.24998912  -1.2499456   -1.24998912  -1.2499456  -10.
 ↪24998912
   -10.24998912]
 [ -0.4         -1.08        -0.4          3.          -9.4
    -9.4        ]]
```

** There are four designated locations in the grid world indicated by R(ed), B(lue), G(reen), and Y(ellow). When the episode starts, the taxi starts off at a random square and the passenger is at a random location. The taxi drives to the passenger's location, picks up the passenger, drives to the passenger's destination (another one of the four specified locations), and then drops off the passenger. Once the passenger is dropped off, the episode ends. There are 500 discrete states since there are 25 taxi positions, 5 possible locations of the passenger (including the case when the passenger is the taxi), and 4 destination locations. Actions: There are 6 discrete deterministic actions: **

```
0: move south
1: move north
2: move east
3: move west
4: pickup passenger
5: dropoff passenger
```

Rewards: There is a reward of -1 for each action and an additional reward of $+20$ for delivering the passenger. There is a reward of -10 for executing actions "pickup" and "dropoff" illegally. Rendering:

```
blue: passenger
magenta: destination
yellow: empty taxi
green: full taxi
other letters: locations
```

State space is represented by: (taxi_row, taxi_col, passenger_location, destination).

Here, the highest number in the array represents the action that the Taxi agent would take.

```
# At state 499 i will definitely move west
state = 499
print(training_memory[0][state])
print(training_memory[20][state])
print(training_memory[50][state])
print(training_memory[200][state])
```

```
[-1.008      -1.0682761 -1.1004      2.72055    -9.2274      -9.1   ⌐
 ↪   ]
[-0.40000039 -1.07648283 -0.40000128  3.          -9.39958914 -
 ↪9.39998055]
[-0.4   -1.08 -0.4    3.    -9.4  -9.4 ]
[-0.4   -1.08 -0.4    3.    -9.4  -9.4 ]
```

```
# At state 77 i will definitely move east
state = 77
print(training_memory[0][state])
print(training_memory[20][state])
print(training_memory[50][state])
print(training_memory[200][state])
```

```
[-1.07999095 -1.008        3.          -1.08309178 -9.1          -
 ↪9.18424273]
[-1.08 -0.4    3.    -1.08 -9.4  -9.4 ]
[-1.08 -0.4    3.    -1.08 -9.4  -9.4 ]
[-1.08 -0.4    3.    -1.08 -9.4  -9.4 ]
```

```
# To show that at state 393, how the move evolved
from IPython.display import Markdown, display
def printmd(string):
    display(Markdown(string))
```

```
action_dict = {0:  "move south"
,1: "move north"
```

(continues on next page)

(continued from previous page)

```
,2: "move east"
,3: "move west"
,4: "pickup passenger"
,5: "dropoff passenger"
}

ENV_STATE = env.reset()
print(env.render(mode='ansi'))
state_memory = [i[ENV_STATE] for i in training_memory]
printmd("For state **{}**".format(ENV_STATE))
for step, i in enumerate(state_memory):

    if step % 200==0:
        choice = np.argmax(i)
        printmd("for episode in {}, q table action is {} and↵
↪it will ... **{}**".format(step*100, choice, action_
↪dict[choice]))
        print(i)
        print()
```

```
+---------+
|R: | : :G|
| : | : : |
| : : : : |
| | : | : |
|Y| : |B: |
+---------+
```

For state **369**
For episode in 0, q table action is 0 and it will ... **move south**.

```
[ -1.24999822  -1.24999945  -1.24999867  -1.24999849 -10.
↪22355492
 -10.24977275]
```

For episode in 20000, q table action is 1 and it will ... **move north**.

```
[ -1.25  -1.25  -1.25  -1.25 -10.25 -10.25]
```

For episode in 40000, q table action is 1 and it will ... **move north**.

```
[ -1.25  -1.25  -1.25  -1.25 -10.25 -10.25]
```

20.3 Running a Trained Taxi

This is a clearer view of the transition between states and the reward that will be received. Notice that, as the reward is consistently high for a trained model.

```python
import time
def print_frames(frames):
    for i, frame in enumerate(frames):
        clear_output(wait=True)
        print(frame['frame'])
        print(f"Episode: {frame['episode']}")
        print(f"Timestep: {i + 1}")
        print(f"State: {frame['state']}")
        print(f"Action: {frame['action']}")
        print(f"Reward: {frame['reward']}")
        time.sleep(0.8)

total_epochs, total_penalties = 0, 0
episodes = 10 # Try 10 rounds
frames = []

for ep in range(episodes):
    state = env.reset()
    epochs, penalties, reward = 0, 0, 0

    done = False

    while not done:
        action = np.argmax(q_table[state]) # deterministic
→(exploit), not stochastic (explore), only explore in training
        env
        state, reward, done, info = env.step(action) #gym

        if reward == -10:
            penalties += 1

        # Put each rendered frame into dict for animation, gym
→generated
        frames.append({
            'frame': env.render(mode='ansi'),
            'episode': ep,
            'state': state,
            'action': action,
            'reward': reward
            }
        )
        epochs += 1

    total_penalties += penalties
    total_epochs += epochs

print_frames(frames)

print(f"Results after {episodes} episodes:")
print(f"Average timesteps per episode: {total_epochs /
→episodes}")
print(f"Average penalties per episode: {total_penalties /
→episodes}")
```

```
+---------+
|R: |  : :G|
|  : |  : : |
|  :  :  :  : |
|  |  :  |  : |
|Y|  : |B: |
+---------+
   (Dropoff)

Episode: 9
Timestep: 123
State: 475
Action: 5
Reward: 20
Results after 10 episodes:
Average timesteps per episode: 12.3
Average penalties per episode: 0.0
```

Here, we looked at how Q-table is being used in Q-learning. However, it is a primitive example as we are dealing with finite states. For infinite states, we would have to rely on a deep learning model instead of a table. This is called Deep Q-learning, which is not covered here.

Bibliography

1. Brownlee J (2019) A gentle introduction to generative adversarial networks (gans). In: Machine learning mastery. https://machinelearningmastery.com/what-are-generative-adversarial-networks-gans/. Accessed 8 Oct 2021
2. Chollet F (2017) The keras blog. In: The Keras Blog ATOM. https://blog.keras.io/a-ten-minute-introduction-to-sequence-to-sequence-learning-in-keras.html. Accessed 8 Oct 2021
3. Chonyy (2020) Apriori: Association rule mining in-depth explanation and python implementation. In: Medium. https://towardsdatascience.com/apriori-association-rule-mining-explanation-and-python-implementation-290b42afdfc6. Accessed 8 Oct 2021
4. Dugar P (2021) Attention seq2seq models. In: Medium. https://towardsdatascience.com/day-1-2-attention-seq2seq-models-65df3f49e263. Accessed 8 Oct 2021
5. eastWillow (2016a) Feature detection and description. In: OpenCV. https://opencv24-python-tutorials.readthedocs.io/en/latest/py_tutorials/py_feature2d/py_table_of_contents_feature2d/py_table_of_contents_feature2d.html#py-table-of-content-feature2d. Accessed 8 Oct 2021
6. eastWillow (2016b) Image processing in opencv. In: OpenCV. https://opencv24-python-tutorials.readthedocs.io/en/latest/py_tutorials/py_imgproc/py_table_of_contents_imgproc/py_table_of_contents_imgproc.html. Accessed 8 Oct 2021
7. eastWillow (2016c) Python tutorials. In: OpenCV. https://opencv24-python-tutorials.readthedocs.io/en/latest/py_tutorials/py_tutorials.html. Accessed 8 Oct 2021
8. eugeneh101 (2021) Eugeneh101/advanced-python: Code like a pro! In: GitHub. https://github.com/eugeneh101/Advanced-Python. Accessed 8 Oct 2021
9. Folsom R (2019) But, what exactly is ai? In: Medium. https://towardsdatascience.com/but-what-exactly-is-ai-59454770d39b. Accessed 8 Oct 2021
10. Foundation PS (2021) The python standard library. In: The Python Standard Library - Python 3.10.0 documentation. https://docs.python.org/3/library/. Accessed 8 Oct 2021
11. Hurwitt S (2018) Classification in python with scikit-learn and pandas. In: Stack Abuse. https://stackabuse.com/classification-in-python-with-scikit-learn-and-pandas/. Accessed 8 Oct 2021
12. Inc A (2017) In: Conda. https://docs.conda.io/en/latest/. Accessed 8 Oct 2021
13. Inc A (2021) The world's most popular data science platform. In: Anaconda. https://www.anaconda.com/. Accessed 8 Oct 2021
14. Li S (2017) Solving a simple classification problem with python - fruits lovers' edition. In: Medium. https://towardsdatascience.com/solving-a-simple-classification-problem-with-python-fruits-lovers-edition-d20ab6b071d2. Accessed 8 Oct 2021
15. Madushan D (2017) Introduction to k-means clustering. In: Medium. https://medium.com/@dilekamadushan/introduction-to-k-means-clustering-7c0ebc997e00. Accessed 8 Oct 2021

16. Pietro MD (2021) Machine learning with python: classification (complete tutorial). In: Medium. https://towardsdatascience.com/machine-learning-with-python-classification-complete-tutorial-d2c99dc524ec. Accessed 8 Oct 2021

17. Rother K (2018) Krother advanced python: Examples of advanced python programming techniques. In: GitHub. https://github.com/krother/advanced_python. Accessed 8 Oct 2021

18. Saha S (2018) A comprehensive guide to convolutional neural networks the eli5 way. In: Medium. https://towardsdatascience.com/a-comprehensive-guide-to-convolutional-neural-networks-the-eli5-way-3bd2b1164a53. Accessed 8 Oct 2021

19. Scikit-Learn (2021a) 1. supervised learning. In: scikit. https://scikit-learn.org/stable/supervised_learning.html#supervised-learning. Accessed 8 Oct 2021

20. Scikit-Learn (2021b) 1.1. linear models. In: scikit. https://scikit-learn.org/stable/modules/linear_model.html#ordinary-least-squares. Accessed 8 Oct 2021

21. Scikit-Learn (2021c) 1.10. decision trees. In: scikit. https://scikit-learn.org/stable/modules/tree.html#classification. Accessed 8 Oct 2021

22. Scikit-Learn (2021d) 1.11. ensemble methods. In: scikit. https://scikit-learn.org/stable/modules/ensemble.html#forests-of-randomized-trees. Accessed 8 Oct 2021

23. Scikit-Learn (2021e) 1.11. ensemble methods. In: scikit. https://scikit-learn.org/stable/modules/ensemble.html#gradient-tree-boosting. Accessed 8 Oct 2021

24. Scikit-Learn (2021f) 1.13. feature selection. In: scikit. https://scikit-learn.org/stable/modules/feature_selection.html#feature-selection-as-part-of-a-pipeline. Accessed 8 Oct 2021

25. Scikit-Learn (2021g) 1.13. feature selection. In: scikit. https://scikit-learn.org/stable/modules/feature_selection.html#removing-features-with-low-variance. Accessed 8 Oct 2021

26. Scikit-Learn (2021h) 1.17. neural network models (supervised). In: scikit. https://scikit-learn.org/stable/modules/neural_networks_supervised.html#multi-layer-perceptron. Accessed 8 Oct 2021

27. Scikit-Learn (2021i) 1.4. support vector machines. In: scikit. https://scikit-learn.org/stable/modules/svm.html#classification. Accessed 8 Oct 2021

28. Scikit-Learn (2021j) 1.4. support vector machines. In: scikit. https://scikit-learn.org/stable/modules/svm.html#regression. Accessed 8 Oct 2021

29. Singh P (2020) Seq2seq model: Understand seq2seq model architecture. In: Analytics Vidhya. https://www.analyticsvidhya.com/blog/2020/08/a-simple-introduction-to-sequence-to-sequence-models/. Accessed 8 Oct 2021

30. Stojiljkovic M (2021) Linear regression in python. In: Real Python. https://realpython.com/linear-regression-in-python/. Accessed 8 Oct 2021

31. Tensorflow (2021a) Basic classification: Classify images of clothing : Tensorflow core. In: TensorFlow. https://www.tensorflow.org/tutorials/keras/classification. Accessed 8 Oct 2021

32. Tensorflow (2021b) Basic regression: predict fuel efficiency: tensorflow core. In: TensorFlow. https://www.tensorflow.org/tutorials/keras/regression. Accessed 8 Oct 2021

33. Tensorflow (2021c) Basic text classification: tensorflow core. In: TensorFlow. https://www.tensorflow.org/tutorials/keras/text_classification. Accessed 8 Oct 2021

34. Tensorflow (2021d) Classification on imbalanced data: tensorflow core. In: TensorFlow. https://www.tensorflow.org/tutorials/structured_data/imbalanced_data. Accessed 8 Oct 2021

35. Tensorflow (2021e) Deep convolutional generative adversarial network: Tensorflow core. In: TensorFlow. https://www.tensorflow.org/tutorials/generative/dcgan. Accessed 8 Oct 2021

36. Tensorflow (2021f) Neural style transfer: Tensorflow core. In: TensorFlow. https://www.tensorflow.org/tutorials/generative/style_transfer. Accessed 8 Oct 2021

37. Tensorflow (2021g) Overfit and underfit: tensorflow core. In: TensorFlow. https://www.tensorflow.org/tutorials/keras/overfit_and_underfit. Accessed 8 Oct 2021

38. Teoh TT (2021) Ai model. In: Dropbox. https://www.dropbox.com/sh/qi7lg59s3cal94q/AACQqaB6mGVSz13PD42oiFzia/AI%20Model?dl=0&subfolder_nav_tracking=1. Accessed 8 Oct 2021

39. vansjaliya M (2020) Market basket analysis using association rule-mining. In: Medium. https://medium.com/analytics-vidhya/market-basket-analysis-using-association-rule-mining-64b4f2ae78cb. Accessed 8 Oct 2021

40. Whittle M (2021) Shop order analysis in python. In: Medium. https://towardsdatascience.com/tagged/association-rule?p=ff13615404e0. Accessed 8 Oct 2021
41. Yin L (2019) A summary of neural network layers. In: Medium. https://medium.com/machine-learning-for-li/different-convolutional-layers-43dc146f4d0e. Accessed 8 Oct 2021

Index

© The Author(s), under exclusive license to Springer Nature Singapore Pte Ltd. 2022   333
T. T. Teoh, Z. Rong, *Artificial Intelligence with Python,*
Machine Learning: Foundations, Methodologies, and Applications,
https://doi.org/10.1007/978-981-16-8615-3

Printed in the United States
by Baker & Taylor Publisher Services